MATTERS OF THE HEART

Frontispiece: 'Heart in her mouth'.
Source: Marion Kaplan.

I drew very near him, my Heart in my Mouth for fear.

Charles de Fieux Mouhy, *The Fortunate Country Maid*
(London: C. Hitch, 1758), 136

MATTERS OF THE HEART

HISTORY, MEDICINE, AND EMOTION

FAY BOUND ALBERTI

OXFORD
UNIVERSITY PRESS

Great Clarendon Street, Oxford OX2 6DP

Oxford University Press is a department of the University of Oxford.
It furthers the University's objective of excellence in research, scholarship,
and education by publishing worldwide in

Oxford New York

Auckland Cape Town Dar es Salaam Hong Kong Karachi
Kuala Lumpur Madrid Melbourne Mexico City Nairobi
New Delhi Shanghai Taipei Toronto

With offices in

Argentina Austria Brazil Chile Czech Republic France Greece
Guatemala Hungary Italy Japan Poland Portugal Singapore
South Korea Switzerland Thailand Turkey Ukraine Vietnam

Oxford is a registered trade mark of Oxford University Press
in the UK and in certain other countries

Published in the United States
by Oxford University Press Inc., New York

British Library Cataloguing in Publication Data

Data available

Library of Congress Cataloging in Publication Data

Alberti, Fay Bound, 1971–
Matters of the heart : history, medicine, and emotion / Fay Bound Alberti.
p. cm.
Includes bibliographical references and index.
ISBN 978–0–19–954097–6 (hbk. : alk. paper) 1. Heart—Diseases—Social
aspects—History. 2. Heart—Anatomy. 3. Heart—Symbolic aspects. 4. Emotions.
5. Mind and body. I. Title.
RC682.A425 2009
616.1′2—dc22

Typeset by Laserwords Private Limited, Chennai, India
Printed in Great Britain
on acid-free paper by
MPG Books Group, Bodmin and King's Lynn

ISBN 978–0–19–954097–6

1 3 5 7 9 10 8 6 4 2

For Millie Bound and Jacob George Alberti

Acknowledgements

A Wellcome Trust Project Grant made possible the research and writing of this book. I gratefully acknowledge the Trust's financial support, and the diligence of the grants administration team, especially Tony Woods. I am also grateful to those who have commented on the ideas and themes expressed in this book. An enormous intellectual debt is owed to Mark Jenner, who supervised my Masters and Doctoral work at York. Other intellectual debts include, at the Wellcome Trust Centre at UCL, Rhodri Hayward, Christopher Lawrence, and Roger Cooter, whose astute criticism was crucial to the development of this monograph. Hal Cook and Anne Hardy have been generous in their time and support, as ever. At the University of Lancaster, I benefited from the expertise of Paolo Palladino and Thomas Dixon, and at Manchester University, from that of John Pickstone and Mick Worboys.

In the medical sphere, George Alberti and Martin Cowie have offered insightful commentary at various stages, both about medical practice and about the relationship between patients and practitioners. In that which is more commonly accepted as a realm of the imagination, I am grateful for the existence of Louisa Young's enchanting *Book of the Heart*, and to Louisa for sharing her own enthusiasm on the subject when we collaborated on the Wellcome Trust's exhibition and book on the heart in 2006. James Peto and Emily Jo Sargent at the Wellcome Trust were also a pleasure to work with. I am grateful to Otniel Dror for sharing with me his insights into twentieth-century medical research, and to Alan Champion, who has worked extensively on the history of Ventnor. I am also appreciative of the support and assistance of the librarians at the Wellcome Library and the British Library.

In researching *Matters of the Heart*, I have spent much time thinking about the relationship between mind and matter, and between the body, the soul, and the brain, in theory and practice. Sam has listened to and shared in many such conversations, and been a generous and scrupulous proofreader. He has also offered challenging scrutiny at many points, especially into the history of pathological bodies. I am very grateful for his help, and for that of my family and friends more generally. Paddy, Nikki, Steve, Lesley, and Matt have each sustained me with babysitting, noodles, or companionship during the times when I thought that the heart could not, after all, 'go on'.

'The heart' as symbolic entity, as organ of the self, is fickle, random, and apparently uncontrollable. Yet 'the heart' as object is circumscribed, controllable—even replaceable. This apparent contradiction still lies at the heart of the organ, and of this book. Its function as a simile, or as metaphor, remains unrivalled. It is, therefore, to the rulers of my own heart that this book is dedicated: to my children Millie and Jacob, whose own heartbeats I have wondered at, fretted over, and been grateful for many more times than I can express.

Contents

List of Figures

Introduction: The Heart of the Matter

> It is as if out of the death of another, new sustenance has been given to the recipient of a transplanted heart. Would the dead person be envious? Would the beneficiary suffer frightening moments of imaginative confrontation with an accusing finger from the person who left him his heart? A certain biblical line from the *Song of Solomon* might assume a new haunting meaning. That dead person who donated a heart might cry with that famous line, 'I sleep, but my heart waketh' (5: 2).
>
> Ali Mazrui, 'The Poetics of a Transplanted Heart'[1]

So wrote the academic and political writer Ali Mazrui in an article entitled 'The Poetics of a Transplanted Heart', published in 1968. In the aftermath of Christiaan Barnard's first successful heart transplant, Mazrui highlighted the peculiar sensibilities attached to the heart as an organ and symbol, and the moral and ethical issues confronting surgeons, donors, and recipients. Aside from the racial questions thrown up by this new field of medical endeavour—in particular the possibility that African people could become 'spare parts for whites'—might it be possible, Mazrui wondered, for the 'borrowed heart' to bring to the recipient something other than a mere pump: a sense of identity, perhaps, even a change of soul?[2]

This question remains with us. Since the 1960s, developments in heart-transplantation techniques have been paralleled by studies suggesting that some emotional or psychological characteristics were transmitted between donor and recipient. In one recent

autobiography, entitled *A Change of Heart*, Claire Sylvia records her
transition from a healthy, active dancer to a heart-transplant patient—a
transition that altered her physical and mental make-up and left her
craving chicken nuggets, a particular favourite of her organ donor that
Claire had apparently never desired before the transplant.[3]

These kinds of stories have been mythologized in popular consciousness. There are a variety of explanations one could put forward
to dispute, or endorse, their legitimacy. For some neuroscientists, it
provides evidence of cellular memory, the organs of an individual
retaining an imprint of lived experience even after brain death. For
others, it is nothing more than a wishful imagining on the part of the
donor family that something of the deceased lives on, albeit in another
breast. For the recipient, it is possible that the sense of 'otherness'
related to the possession of another's heart gives rise to emotional
trauma and subsequent psychological alteration. The status of such
competing narratives as 'truth claims' is ultimately unimportant. What
is significant from the point of view of the cultural history of the heart
is that such questions exist. What is it about the heart as an organ that
has inspired such tales of personality transfer? After all, there seem to
be no parallel claims of kidney recipients waking up to find themselves
classical pianists.[4]

Emotions, the heart, and the 'self' (however defined) have been
linked in medical and popular consciousness for many centuries. We
'feel' in the heart. We 'know' from the heart. Sometimes we privilege
our 'hearts' over our 'heads', claiming emotional over rational reasoning as the appropriate response to any given circumstance. Or person.
But where does this language come from? And what does it mean?
Furthermore, how do such tales of personality exchange continue
to thrive when, to most physicians, the heart is nothing more than
a pump? *The Oxford English Dictionary* provides a definition of the
'heart' (n.) that is common to most modern works of medical reference: 'The hollow muscular or otherwise contractile organ which, by
its dilatation and contraction, keeps up the circulation of the blood in
the vascular system of an animal.'[5] That seems straightforward enough.
And yet underneath that definition is the following alternative, but
'archaic', definition: '2. Considered as the centre of vital functions: the
seat of life; the vital part or principle; hence in some phrases = life.'
And even more surprisingly, perhaps: '5.a. MIND, in the widest sense,
including the functions of feeling, volition, and intellect.'[6] These are

rather a lot of definitions for one organ to acquire. They are also rather contrary definitions.

This apparent divergence between the heart-as-pump found in science, and the heart-as-emotion rhetoric that survives in popular culture has seldom been explored. This is arguably because of the ordering of disciplinary categories—cardiology (the modern study of the structure and functions of the heart) being structured and studied in isolation from the mind sciences, which focus on the study of mental and behavioural characteristics. The study of emotions has continued a separate trajectory, whether in the mind sciences, the social sciences, or more recently the humanities, as part of the history of the secularized mind or self. There was little research into the relationship between the heart and emotions during the eighteenth and nineteenth centuries, a crucial chronological point in the solidification of many of the categories (mind, body, heart, emotions), in common usage in the twenty-first century.

To account for the diverse languages of the heart, and to answer some of the questions outlined above, we need to explore its medical and cultural history. It is only by unravelling the theorizing of the heart, from the classical world to the twentieth century, that we can understand the lingering language and sentiment behind the heart's emotional status. Yet such an enterprise throws up still more questions: how does the heart link to the brain? Where is the self located? And how (and when) did emotions (along with concepts of 'the self') become redefined as mental, rather than bodily, experiences? According to the famous and oft-quoted Cartesian formulation, *cogito ego sum* (I think, therefore I am). We might 'feel' in the heart, but those feelings originate in thought and cognition, the essence of humanity being located in the brain. That, at least, is the 'common-sense' modern approach. Why, then, does the heart continue to hold such resonance as a feeling organ, most associated with love (I ♥ U); with courage (the heart of a lion); and with truth (because my heart tells me so)? In addressing these themes, this book presents the following arguments.

First, the cultural and spiritual origins of the heart as a symbol of affect (and affection), and as an organ of emotion, were embedded in classical ideas about the body and the mind, and retained and transmitted into Western theory for thousands of years by the Galenic medical tradition. Many of the 'heartfelt' languages and images of the heart and the blood that remain with us, such as 'cold hearted', 'warm hearted', 'hot blooded', and 'cold blooded', derive from Galenic principles. These concepts have lingered as (possibly self-reinforcing) metaphors where once they were taken as a series of bald, medical facts. The heart warmed the blood in order to generate and sustain particular emotional states; it moved in response to the sensations of anger, love, and fear; it was affected by the operation of the soul, with its links to mind and body.[7] These beliefs about the active status of the heart in producing emotions were linked to the physical, lived experience of emotions themselves: passions were *felt* in the breast, whether in the dull ache associated with the loss of love, or the palpitations brought by excitement.

Second, the historical development of alternative ways of understanding the actions of the heart, be they mechanical, chemical, nervous, or hormonal, took place in the context of attempts to redefine and understand the human body and its links to the soul and the divine. The dominance of any one narrative of human physiology was not inevitable; incompatible stories often nestled curiously side by side. One obvious example is the theory of blood circulation as espoused by William Harvey in the seventeenth century (in a discovery that borrowed heavily from earlier Islamic writings on pulmonary circulation, notably by Ibn al Nafis), but that did not radically alter medical interpretations or practices, most notably bloodletting, for many years to come.[8] This curious phenomenon is an example of the disjointedness of theory and practice in the fields of science and medicine, a factor that must be borne in mind when we consider changing perspectives on the heart and its functions.

Third, medical conceptions of the heart as a pump (responsible for the circulation of the blood and as subject to decay, as any other material structure) came into being only after the theorizing that followed Harvey's discussion of blood circulation. As a matter of interest, the 'pump' metaphor was made possible by the contemporaneous popularization of mechanical pumps of all kinds—a process that is indicative of the analogical process by which human emotions have been historically understood.[9] The language in which the mind–body–brain relationship has been addressed throughout history is therefore indicative of broader cultural shifts in social and economic life, as well as the available metaphors to articulate difference. In the same way that metaphors of clockwork bodies (and

hearts) became conceivable at the same time as manufacturing mechanisms made such phenomena part of the material world, there has
been a post-twentieth-century shift towards seeing the mind as a
complex neuro-physiological structure akin to a computer; memories
and experiences exist in separate compartments or files, to be accessed
when required by the mainframe operator. Other examples include
the mind as a 'filing cabinet' and the hydraulic body that boils, like a
kettle, unless emotions are released.[10]

From the late seventeenth century, the heart was dissected, examined, labelled, and catalogued in a way that was hitherto unknown.
Its functions were exposed to the processes of morbid anatomy by
anatomists like the brothers William and John Hunter. How ironic,
then, that it should be John Hunter's heart that provides us with the
first case study in this book.[11] At his death in 1793, the surgeon's
heart was subjected to the same classificatory principles of the organs
lining his museum shelves: its thickened, calcified structure provided
evidence of 'angina pectoris', one of several 'new' cardiac diseases to
come into being from the late eighteenth century, and as much a
product of emotional distress as structural disease.

The links with emotion and heart disease are important. The fourth
argument presented here is that a crucial aspect of eighteenth-century
classifications of heart disease was a reformulation of the role, and
the extent, of emotional influence. Traditionally, emotions were
perceived capable of causing profound structural changes in the body,
and in the mind. This was a throwback to classical beliefs that the
passions, working in conjunction with the humours to effect the
soul's desires, could physically alter the constitution of the human
body.[12] By the nineteenth century, the belief in physical (or *structural*)
alteration was limited, as cardiac phenomena began to be explained
in terms of nervous transmission. Far more likely, it was believed,
emotions could cause a disruption in the operation of the heart (a
so-called *functional* disease), perhaps as a result of a disruption in the
circulation. It was possible that functional and structural disease could
be related (repeated functional problems causing actual changes in the
structure of the heart), but emotions gradually became associated with
functional disease (and with women) through concepts of nervousness
and neuroses.[13]

This shift in interpreting the relationship between structure and
function (and the gendering of that relationship) takes us to the fifth

argument presented in this book. By the time of Thomas Arnold's death from angina pectoris in 1842, it was common for angina pectoris to be regarded as a structural disease when associated with men of a particular weight, lifestyle, constitution, and personality.[14] It was also increasingly commonplace (though this would not peak until the early twentieth century) to associate functional angina pectoris with women. Beyond the gendering of angina pectoris, however, many more forms of heart disturbance were being regarded as functional by the end of the nineteenth century. This claim runs counter to established historical writings about the trajectory of heart disease from functional to structural as new forms of disease causation were identified.[15]

The transition of angina pectoris from a largely structural to a largely functional disease was made possible through a simultaneous reconsideration of the parameters of 'the emotional', a theme that this book explores in some detail.[16] This claim is consistent with the development of an ostensibly modern, scientific, and rational medicine from the nineteenth century as part of the process of modernity—the development of institutions and authorities that formalized the statuses of health and disease, promising also transformations in understandings of interiority, human behaviour, and the self.[17] Fernando Vidal has recently made a similar claim about the status of the modern self, stating that as a 'cerebral subject' a human being is defined 'by the property of "brainhood", i.e. the property or quality of *being*, rather than simply having a brain'.[18] Vidal's argument is supported in this book by an examination of the rise of the self and the emotions as products of mind in late-nineteenth-century culture.[19]

Vidal's discussion of the birth of the cerebral subject is unusual in that it does not seek to identify a parallel emergence of self at the level of practice. Historical discussions typically, and problematically, focus on experiential changes, most often through an over-reliance on Cartesian principles (often taken out of context and/or oversimplified), and the study of the language of emotions (what Peter and Carol Stearns termed 'emotionology') as being tantamount to studying emotions themselves.[20] Nevertheless, it is interesting to view the construction of the modern heart, the heart of science, perhaps, in the context of the rise of modernity and the emergence of the subjective self as a unit of discourse, if nothing else.

It can be no coincidence that, at the same time as scientific theorizing of the body as a set of separate if interrelating systems was taking place (a separation that would be embedded in the construction of laboratory experimentation as well as in the newly emergent hospital system), concepts of 'the mind' and the emotional were also subject to scrutiny. Moreover, the rise of the mind sciences and attempts to define and classify a 'science of emotion' took place to mirror the rise of experimental physiology.[21] Debates centred on how far emotions existed beyond their physical manifestation, and how far they constituted cognitive processes that could be detached from the physical realms. Part of the process by which this degree of experimentalism occurred was the rise of the mind sciences and the evolution of the 'feeling' brain, a shift from cardio-centrism to cranio-centrism that remains entrenched in twenty-first-century medical theorizing.[22]

Finally, therefore, this book will suggest that the transition from heart to brain at the level of theory took place under the influence of processes that we might broadly associate with modernity. The construction of the subject *as* a thinking being, the cognitive processes located in the brain (rather than in the soul), and not in the body, meant that it was the brain that would be prioritized in discussions of emotions and the self. Since the nineteenth century and the segregation of the body and the mind into a series of disparate parts, it has become commonplace for the brain to be the organ most associated with life (indeed, 'brain death' has succeeded 'heart death' as the number one criterion for determining life cessation).[23] Psychology, psychiatry, and neuroscience each posited the brain (that cold, wet matter that was all but dismissed by Aristotle) as the site where intelligence, emotion, thought, and the self originated.[24]

The institutions and procedures that catered for this newly circumscribed self were attended by objectifying measures that removed the need for subjective assessment, placing interpretation in the hands of a trained (and usually male) expert. The technologizing of medical knowledge by diagnostic aids and practices from the late nineteenth century created a hierarchy of knowledge in which individual experiences of cardiac sensation were downplayed in favour of objective measurement. As the epigraph suggests, however, things were not so straightforward. The shift in theory—away from the heart of feeling and towards the brain of feeling, away from the heart as symbol and towards the heart as organ—was ultimately less successful at the level of practice. In a mid-nineteenth-century case study, we find that Peter Mere Latham, or 'Heart Latham', continued to treat his patients in the way medical practitioners had long done—as holistic entities, whose hearts beat in correspondence with their general constitution and lifestyle.[25] The century saw a number of important diagnostic strategies and technological innovations designed to 'know' the living heart, and to standardize that knowledge in order to construct new theories of disease. Yet the preservation of holism and the reluctance of many physicians to incorporate those innovations in a clinical setting meant that genuine medical specialization was unusual even in early twentieth-century Britain.[26]

Moreover, at the same time as the heart became more materially ordered and circumscribed in scientific discourse, its status as a cultural artefact became paradoxically more emotional: the feeling heart was crucial to the Romantic project and provided evidence of creativity and the divine.[27] This phenomenon draws attention to further ambiguities in relation to the construction of emotions as mental phenomena. Although it has been a part of modernity's project (intentional or otherwise) to separate the mind and the body, to prioritize the mind (the masculine, the cerebral, and the rational) at the expense of the heart (the feminine, the somatic, and the irrational), it remains the brain that has a hold on emotions as part of its cognitive capacity. Emotions are felt by the body, and yet belong to the mind; the heart is a mere respondent to the sensations and experiences being cognitively processed by the brain.

This realization points to a crisis at the heart of modernity and the feeling subject. For the heart cannot be both pump and feeling organ under scientific accounts of the mind–body relation. The preservation of 'common-sense' ideas grounded in individual and collective experience of cardiac phenomena (as seen in the disputes over heart transplantation) has opened up new spaces of meaning where the heart has been reaffirmed as an intelligent organ, a subtle repository of the self. The scientific community has responded in two main and very different ways; first, by rejecting the existence of any emotional status being attached to the heart, or, alternatively, by incorporating and transforming these counter-narratives. In the case of the latter, new languages of emotional intelligence grounded in materiality and scientific discourse are being used to redefine and rework emotional memories. These include concepts of 'cellular memories', and of the heart possessing a 'little brain', or a pathway of 'neurons and synapses' akin to those that exist in the brain.[28]

Who knows where these reformulations will take us. What is interesting is that they are necessary in the West, where scientific medicine has become dominant, and less so in traditional Eastern medicines, where the brain and the heart are regarded as indivisible.[29] Will the reworking of the heart and brain by organizations like 'Heartmath' reunite the mind and body in a material framework and take us closer to the humours than the hormones did? Will the 'little brain in the heart' gain stature and relevance as time progresses, perhaps even overtaking the 'wet cold matter' of the brain? Or will the brain and mind be mapped in radically new ways that reassert the inherently physiological role of the heart?[30]

These kinds of issues are obviously beyond the scope of this project. But they need to be acknowledged if we are to understand how important the heart's history is to modern debates about emotions, the brain, and the mind–body relation. As theories of the body, of identity and the self, come under increased scrutiny—through such controversial issues as the Human Genome Project, retained organs, stem-cell research, and animal-human hybrids—the history of the heart has never seemed so relevant.[31]

Emotion History and Terminology

Since the history of emotions is a rapidly developing field, its own history and terminology need to be acknowledged.[32] Developing from social and cultural history's interest in the 'personal', or the 'psychological', 'emotion history' has become a discreet subdiscipline in its own right, from the ground-breaking work by Lucien Febvre in the 1940s, through a series of works by Carol and Peter Stearns in the 1980s and 1990s, that prioritizes the question of emotional change, and its implications for subjectivity, interiority, and selfhood.[33] More recently, historians from diverse chronological periods have explored specific emotions, such as anger, jealousy, and fear.[34] Interestingly, the emotions that are chosen for research are those that have somehow been linked to modern subjective and individualistic identities: there is no similar outpouring of works on 'collective' imaginings and emotions, whether linked to benevolence, compassion, or mob rule.[35]

Much emotion history has been about language; from the 'emotionology' of the Stearns to the 'worrying' of Rosenwein, historians have attempted to explain the different emotional climates (or 'communities') prevailing in particular epochs.[36] Other challenges that unite historians include the themes of rhetoric, cultural specificity, and the connections between what philosophers have called the 'raw feels' (bodily sensations), and cognitive experiences of emotion, all of which concerns are shared by scholars in anthropology, sociology, and psychology.[37] Rather less attention has been paid to embodied emotions, or to the physical, lived realities of feelings. This is perhaps the most difficult aspect of emotions to address. Even if we suppose that emotions exist as bodily artefacts outside of language, how do we begin to access them? One answer, at least for twentieth-century scientists, comes from the measurement and quantification of physiological indices: from raised heartbeats to cold sweats.[38] As will be seen, however, there is no easy correlation between emotional experience and its expression; making the heartbeat into an autonomic response dependent on a number of physiological variables potentially reduced its ability to convey psychological experience.[39]

Rather than providing a comparative study of emotions over time, or a linguistic account of the shifting emotional states associated with the heart, this book explores the shifting status of the heart and its links with affective states.[40] Its focus is less on the lived experience of cardiac sensations (though with an alternative source set that would prove a valuable study) than with the relationship between medical and constructions of affect as linked to the heart as an organ. Since most discussion on the heart of emotion in medical writings focuses on its pathology, there is a source-led emphasis in this book on the heart of disease—the heart that was taken apart, dissected and studied to find out what it could reveal about the role of extreme emotional states.

The heart—diseased or healthy—is more neglected as an aspect of the history of emotions than as a cultural or medical artefact. There is an extensive history of cardiology that traces the emergence of medical specialization by the early twentieth century, often stretching its origins back into the seventeenth century and beyond.[41] There has also been much work on the history and origins of cardiac surgery, which incorporates the history of heart transplantation.[42] Medical historians are seldom concerned with the heart as an emotional organ, but there have been some important studies on the heart as symbol that include historical and/or medical context. Louisa Young's *Book of the Heart*, for instance, provides a rich and diverse account of the cultural embeddedness of heart iconography throughout a range of times and cultures. More recently, and with relevance to the history of the heart as a medical organ, Kirstie Blair has studied the emergence of the pathological heart in Victorian literature in an insightful work that reminds us of the embeddedness of fictional and non-fictional texts in the pursuit of knowledge.[43]

Throughout this book, there is an awareness and an acknowledge-ment of the breadth of the heart as an object of study, its material, emotional, and symbolic properties viewed in many different ways, in different times, and in different contexts. The concept of 'heart' is more difficult to pin down than its material, physical incarnation as an object of science. The same must be said about emotions, as transitory experiences that are notoriously difficult to define or to structure. Yet deconstructing emotional codes tells us much about the society in which they function. In a recent series of essays by the Polish linguist Anna Wierzbicka, for instance, emotions are shown to be communicated and structured as linguistic as well as bodily experiences (through scripts of 'sincerity' or 'warmth') that reveal much about culturally situated values and norms.[44]

Given the complexities of the subject matter, then, it is clearly necessary to say something about the terminology employed in this book. The languages of emotion used at different epochs, and in different sources, have varied considerably. It is often difficult to distinguish expressed codes of emotion from their specific sites and circumstances of production. This is as true for seventeenth-century court records and emotional scripts as it is for the language of marital disputes, love letters, and narratives of illness and suffering.[45] Different spaces of textual production of emotions, in the past as in the present, depended on a variety of different registers.

The languages of emotion are problematic across different historical periods, as well as across cultures and genres. The 'passions' of the seventeenth century (with all their religious overtones of suffering and the Passion of Christ) cannot be unproblematically aligned with emotions in the eighteenth and nineteenth centuries. In a recent book

on this subject, Thomas Dixon has identified a linguistic shift between
the earlier and the later period, a shift that arguably reflects broader
transformations in philosophical and theological attitudes towards the
status of emotions as human artefacts.[46]

There have also been historic shifts in the numbers and types of
emotions that are classed *as* emotions. Part of the reason why early
modern attitudes seem so different from their modern counterparts is
the number of 'passions' they also described as 'inclinations', 'pertur-
bations', or 'yearnings'. Under humoral theory, passions and emotions
were both psychological and physical states that worked on the mind,
the body, and the soul. Modern-day distinctions between mental and
cognitive *or* physical and somatic experiences had no relevance. Nevertheless, debates continued about such themes as what constituted an
emotion, how many emotions were peculiar to humans, and whether
they originated in the mind, the soul, or the body. Their impact
was similarly controversial: in medieval and early modern Europe,
emotions were said to be both beneficial and detrimental for the
health, linked to the bodily and the secular, but also to the spiritual
and divine realms.[47] They were God-given, yet (in some cases) shared
with animals; they were associated with ethical principles, for good
and ill. How far emotions could and should be controlled, acknowledged, denied, or uncovered has been a staple of debates over health
and disease from the medieval period to the present.[48] So, too, have
discussions of human motivation, one characteristic of modernity
being that the motive behind emotions has shifted from the soul to
the individual psyche. Even in the twenty-first century there remains
uncertainty and conflict about the numbers or types of emotions, as
well as about their purpose and development.[49]

In focusing explicitly on the organ of the heart and its links to
emotion, this book acknowledges the social and cultural specificity of
beliefs about affect. With the exception of seventeenth-century texts
that explicitly use the term 'passion', this book uses the term 'emotion'
to mean physiological experiences with a cognitive dimension—that
is, that were recognized as distinct and recognizable psychological
experiences by eighteenth- and nineteenth-century writers. Each of
the chapters below follows the terminology used in the primary
sources consulted, and there is no attempt to equate the experiences we might recognize in the twenty-first century (such as
'stress') with those prevailing in earlier periods.[50] Finally, this book

chooses not to engage with theological or philosophical discussions about passions, ethics, and moral philosophy, on the grounds that this opens up too many registers of emotion rhetoric. Other scholars have addressed these themes in considerable detail, and there is simply insufficient space to include them here.[51] It would be interesting to trace further the connections between morality, ethics, and the heart as identified under Romanticism, however, or to analyse the shifting status of cardiac function as related to theories of life and death.[52]

In employing the same terminology as contemporary patients and practitioners, this book traces medical and cultural understandings of emotions as both physical and psychological events, exploring the extent to which transformations in scientific theory impacted upon, or fed back into, cultural practices (and vice versa). (One such example might be the way that medical ideas about palpitations and cardiac function indicating emotional sensitivity moved between medical and literary writing about the heart.[53]) In so doing, it explores and evaluates the status and meanings of the emotional heart between the seventeenth and nineteenth centuries. Particular emotions that caused concern to early modern commentators were anger and anxiety.[54] Attitudes towards such extreme emotions reveal much about the ways their existence has been understood to impact first on the physical structure of the heart, and second on its functions. Charting the development of medical and cultural beliefs about the interconnectedness of heart, mind, body, and soul between the seventeenth and nineteenth centuries also reveals much about the gender- and class-based analyses at work in the construction of the cardiac patient.

The structure of this book is thematic, and broadly chronological. It begins with a survey of the links between emotions and the heart between the seventeenth and the early twentieth centuries. This is not intended to be exhaustive, or to impose any sort of teleological transition. Rather, it is to provide an episodic outline of some of the major ideological and physiological underpinnings of emotion rhetoric during this *longue durée*. The emotional significance of the heart, both physiologically and spiritually, in the early modern West (particularly in relation to the transmission of classical sentiment), is traced through the hydro-dynamical and chemical theories of the seventeenth century, to the nervous physiology of the eighteenth and nineteenth centuries, and the emergence of modern understandings of nervous influence and the mind-body relation.

The emergence of hormones as an explanatory category by the turn of the twentieth century, and the reintegration of mental and physical processes, take us back, almost full circle, to humoral influence and mind–body holism.[55] And yet there were some very important differences between the theorized heart of the early twentieth century and that of earlier periods. In the former, the heart was no longer the centre of emotions, merely the site where emotions might be felt, and not by the actions of the soul, but by basic biological (often mechanical) processes related to blood pressure and hormonal surges. More fundamentally, moreover, those experiences were not controlled in the heart itself as an organ, its structure being simply muscular. Rather, the controlling organ was the brain. It was in the brain that cognition, apprehension, and awareness now took place; it

Fig. 1. Jennifer Sutton and her dead heart.
Jennifer Sutton, a 23-year-old heart transplant patient, confronting her own dead heart, shown as part of a Wellcome Trust exhibition held in 2006.
Source: Adrian Brooks.

was from the brain that impulses were sent to the heart, and it was through the brain that we became aware of sensations in the body.

By the closing decades of the nineteenth century, this interpretation of the mind-body relationship, and of the heart's subservience to the brain (with the rest of the body acting as agents to undertake the brain's commands), rapidly acquired some 'common-sense' status in the modern West. It arguably represented the triumph of reason over passion, of mind over matter, of masculine over feminine. And yet the heart continues to dominate as an emotional symbol. In popular culture, it is the heart and the heart alone that stands as a cipher for emotion, especially for romantic love. It is also the heart that continues to dominate debates over transplantation. The unsettling image of Jennifer Sutton confronting her own dead heart at the 2006 Wellcome Trust Heart exhibition provides a dramatic illustration of our attitudes towards the heart as an emotional and somehow *personal* organ (Fig. 1.).[56] Given the shifts that have taken place in scientific theorizing since the end of the twentieth century, moreover, it may not be long before the 'heart of emotion' is returned to mainstream science: a heart conceived not merely as an organ or pump for the circulation of blood but, like the brain, as an intelligent organ that retains memories, experiences, and emotions.[57] These are the kinds of issues that this book explores, using matters of the heart to locate emotion in medicine and culture.

I

Humours to Hormones: Emotion and the Heart in History

Heart (n.)

1. a hollow muscular organ that pumps the blood through the circulatory system by rhythmic contraction and dilation.
2. the central, innermost, or vital part of something.
3. a conventional representation of a heart with two equal curves meeting at a point at the bottom and a cusp at the top.
4. (hearts) one of the four suits in a conventional pack of playing cards, denoted by a red figure of such a shape. • a card of this suit.
5. a person's feeling of or capacity for love or compassion. • mood or feeling: they had a change of heart. • courage or enthusiasm: you may lose heart as the work mounts up.
6. the close compact head of a cabbage or lettuce.
7. the condition of agricultural land as regards fertility.

The Concise Oxford English Dictionary (2008)[1]

Tracing the history of the heart of emotion is a complex task. As indicated by the epigraph, there are many different kinds and meanings of 'heart', of which definitions numbers 1 and 5 are the most meaningful for us here. There are many texts about the heart as an organ (category 1), mainly under the categories of physiology and pathology. Since the nineteenth century the size and structure of the heart have been expounded in terms of quantity (such as how many pints of blood the heart can pump around the body each day),[2] and disease (particularly when related to the artery-clogging behaviours of the West).[3] There are also many books on the heart as an emotional structure (category 5), grouped under such headings as poetry, romantic literature, moral philosophy, and the 'sacred heart' of Catholicism.[4] The meanings and history of human emotions, however, have, since the birth of the mind sciences in the nineteenth century, predominately been shelved under psychology, biology, or the neurosciences. These divisions or categories have little to do with the heart and everything to do with the mind, or, more specifically, the brain, reflecting the modern-day emphasis on the brain, rather than the heart, as the organ and site of emotion.[5]

The emergence of emotions under these classifications (and the separation of emotion from the heart in an explanatory sense) tells us much about the historical evolution of the sciences. This is a theme that will be addressed in detail below.[6] For, even allowing for the emergence of 'social-selves' psychology, most mind-science research historically focuses on individualistic models of human emotions, and the individual's mental and cognitive processes as linked to the, often abstract, category of 'mind'.[7] And yet the rise of mind is comparatively recent, as is the identification of the brain as the site of emotions in the human body. Before the nineteenth century it was quite common for emotions to be lumped together, often with the more outmoded term 'passions', with broader concepts of 'moral philosophy' and ethics.[8] General medical practice and literary texts, moreover, both historically recognized emotions as physical *and* psychological events felt in, and symbolized by, the heart.[9] Gradually, however, and with the emergence of scientific medicine (and the separation of mind and body, head and heart), theoretical ideas about the mind and the body transplanted emotions to the mind. And the heart became an organ of the body: mechanized, predictable, subject to decay and the barometer (rather than the instrument) of emotional experiences.

This scientific transition from a cardio-centric to a cranio-centric body is one that has been largely unexplored as an aspect of the histories of medicine, emotions, and the body. Yet there is no way of understanding modern ambivalence towards the heart, as an emotional symbol, on the one hand, and a functional organ, on the other, without understanding the weight and perseverance of the networks of meanings—spiritual, medical, psychological, and physiological—in which it has historically been embedded.

In this chapter, then, I will sketch the status of the heart as an emotional organ between the classical period and the nineteenth century, a broad process of change or transformation that arguably embraces something larger than the heart itself: the concepts of mind, body, and emotions in which the heart was habitually embedded. There is no intention to impose a linear chronology of change; rather, episodic moments in understanding the heart have been singled out for analysis. By putting back together categories 1 and 5 of the epigraph, I will show how and why emotions and the organ of the heart historically evolved as fundamentally connected entities (linguistically, as well as psychological and physiological experiences). I will then suggest, with reference to the shifting status of the mind-body relationship in the West, how it is that heart and emotions were subsequently separated in medical theory in favour of a story of human emotions focusing on the mind and the brain.

Heart and Soul: The Classical Inheritance of Humoralism

Until at least the late seventeenth century, interpretations of human emotion focused on the humoral pathology of classic inheritance, the most eloquent or long-lasting exponent of which was Galen of Pergamum. In the two thousand years after Hippocrates, 'Galenism' became a set of connected principles, doctrines, and concepts that dominated psychological and physiological theories—including emotion beliefs—throughout much of Europe until at least the eighteenth century.[10] Under Galenism the human body was a 'little world' or microcosm of the universe, the tripartite divisions of heaven, sky, and earth corresponding to the three main parts of the human body: the head (reason), the breast (heart), and the lower body (nourishment, procreation).[11] The soma, therefore, possessed all the qualities that made up the 'greater world' of fire, air, water, and earth.[12] Four qualities—hot, cold, moist, and dry—inhered within these elements: fire was hot and dry; air was hot and moist; water was cold and moist; and earth was cold and dry. In each individual, characteristics received the form of 'humours', which were believed to course through the body: blood, which was hot and moist like air; choler (or yellow bile), which was hot and dry like fire; phlegm, which was cold and moist like water; and melancholy (or black bile), which was cold and dry like earth.[13]

The proportional balance of these humours within each individual was partly engendered, and partly innate; a product of heredity, age, sex, and what contemporaries called the six 'non-naturals': air; food and drink; exercise and rest; sleep and waking; evacuation and repletion; and passions of the soul. Although the passions acted on the spirits and humours, therefore, they were also influenced by, and a product of, humoral balance. And an individual's humoral balance was environmental as well as constitutional. As the seventeenth-century English philosopher Thomas Hobbes put it, their proportions 'proceedeth partly from the different constitution of the body, and partly from different Education'.[14] Nevertheless, the overall composition of the humours determined an individual's psychological and emotional demeanour, as humours were produced in the liver and coursed through the veins, mingling with the blood and affecting the mind, soul, and body alike as it did so. A disproportionate amount of any of the humours led to constitutional imbalance, including illness, as well as extreme emotional types, and humours and emotions were inseparably linked. In the words of Wright, 'passions engender humours and humours breed passions'.[15]

Although the language of 'personality types' or 'characters' was not commonly discussed until much later, early modern men and women were understood to display certain psychological tendencies, as expressed in language that we recognize today. A high level of yellow bile made men and women subject to anger (choler) and black bile to sadness (melancholy), whilst an excess of blood or phlegm made one sanguine (and prone to love-sickness) or phlegmatic. Each age and sex was accorded specific 'prevailing humours' as 'the manners of the soul follow the temperature of the body'.[16] As Lelland J. Rather has put it, 'young men are hot, incontinent and bold, old men are cold, covetous and cautious, women are envious, proud and inconstant—and these differences rest on differences in corporeal makeup'.[17]

As this evidence suggests, humoralism was inherently gendered. Emotions and passions were often mixed and interrelated. The English divine and scholar Robert Burton wrote that melancholy could be caused by such diverse passions as sorrow, fear, envy, shame, hatred, anger, ambition, and self-love.[18] Yet there were specific emotional tendencies characteristic of each sex, and those tendencies were apparent in physical differences. Women tended towards a phlegmatic or cold and moist disposition, since their bodies were fleshier, softer, and weaker than those of men, their hair longer, their faces paler, and their skin more moist. The greater passivity of women also made them more subject to such emotional extremes as hysteria.[19] Men, by contrast, with their leaner bodies and drier complexions, tended to display qualities of courage and anger. This model was also generational, with men in particular being prone to varying emotional behaviours during the course of their lives. In the words of Wright, 'younge men generally are arrogant, prowde, prodigall, incontinent', and old men 'subject to sadnesse caused by their coldness of blood'.[20] Moreover, not only were emotional capacities gendered, but also emotional expressions, such as weeping. The preponderance of water in women's physical constitution meant that women were more prone to tears, and also to sudden, irrational rages, since women's flesh 'is loose, soft and tender, so that the choler being kindled, presently speeds all the body over, and causeth a sudden boyling of the blood about the heart'.[21] Women's anger soon passed, however, since (like old men) they lacked the heat to sustain the emotion.

In these discussions of humoral physiology, emotions were grounded or concretized in the physical body and—more specifically—in the organ of the heart. This is crucial to any understanding of early modern emotion physiology, for, while the brain was the seat of reason, the heart was the site of emotion or passion. As Burton put it, this organ was 'the seat and foundation of life, of heat, of spirits, of pulse and respiration, the sun of our body, the king and sole commander of it, the seat and organ of all passions and affections'.[22] In the heat-based economy of emotion physiology, the heart was the centre point where blood was heated or cooled, respectively, under such passions as anger or fear. Compare Pierre de la Primaudaye's conduct-book description of rage, for instance, with the preaching of the English divine John Downame:

For first of all when the heart is offended, the bloud boyleth round about it, and the heart is puffed up: whereupon followeth a continuall panting and trembling of the heart and breast.[23]

[Anger is] an affection, whereby the bloud about the heart being heated, by the apprehension of some injury offered to a man's self or his friends, and that in turn, or in his opinion onely, the appetite is stirred up to take revenge.[24]

This emphasis on the heart as the agent of heating and concoction was compatible with an emotion physiology that linked body, soul, and mind in a complex union. It also derived from a more ancient set of emotion beliefs. Eric Jager has shown how Aristotle (heavily influential after the twelfth century in the West) also associated the heart with the vital functions, emotions, and sensation, whilst classical Latin commonly used the heart (*cor*) 'as a synonym for thought, memory, soul and spirit, as well as for the seat of intelligence, volition, character and the emotions'.[25]

Because Renaissance physicians regarded the soul and the body as indivisible, the soul was necessarily involved in the production of emotions as bodily experiences. As 'operations of the soule, bordering upon reason and sense, prosecuting some good thing, or flying some ill thing [and] causing there withall some alteration in the body', emotions were essentially cognitive phenomena. As Wright continued, they were 'movements' of the appetitive faculty (in the mind/soul) that instigated a 'passion' or a change in the humours.[26] This role accorded to the appetitive faculty gave emotions a moral dimension, as to fear or loathe an object was God-given. And it made the physiological process of the passions and their mediating point between *psyche* and *soma* quite logical. The communication of an image via the senses to the brain preceded any judgement about its value; a corporeal alteration in the brain caused 'spirits' to move to the heart, where they would 'signify' the object, and the heart would bend itself to seek or avoid the same.[27]

The soul, therefore, had 'power to excite Corporeal Passions *directly*, that is, without considering successively various things', as 'is manifest from her sovereignty over the body, which in all voluntary actions is absolute and uncontrollable'.[28] It is in the context of corporeal alteration that the heart was most evident, for in order to implement the desires of the soul the heart required humours: melancholy blood to produce pain and sadness; blood and choler for anger. The humours were concocted in the heart, and the spirits produced were sent around the body to effect the soul's desires. The agitation of the spirits in joy and the free flow of the blood throughout the body were in direct contrast to the physiological experience of fear, when the blood retreated and the soul shrank back from the perceived threat. This process explained the physical manifestations of emotion. As Downame's *Treatise of Anger* continued, the passion:

Maketh the haire to stand on end, shewing the obdurate inflexiblenesse of the minde. The eyes to stare and candle, as though with the Cockatrice they would kill with their lookes. The teeth to gnash like a furious Bore. The face now red, and soon after pale, as if either it blushed for shame of the mind's follie, or envied others good. The tongue to stammer, as being not able to express the rage of the hart. The bloud ready to burst out of the vaines, as though it were affraide to stay in so furious a body. The brest to swell, as being not large enough to containe their anger, and therefore seeketh to ease it selfe, by sending out hot-breathing sighes. The hands to beate the tables and walles, which never offended them. The joyntes to tremble and shake, as if they were afraid of the mines furie. The feete to stamp the guiltlesse earth, as though there were not room enough for it in the whole element of the aire, and therefore sought entrance into the earth also. So that anger deformeth the body from the hayre of the head to the soale of the foote.[29]

If anger caused the blood to boil around the heart, the reverse physiological process was associated with fear, as with grief and sorrow. These 'negative' emotions caused the soul to contract. Thus, in grief and sorrow, 'the *Animal Spirits*' were

recalled inward, but slowly and without violence: so that the *blood* being by degrees destitute of a sufficient influx of them, is transmitted with too slow a motion. Whence the pulse is rendered *little, slow, rare* and *weak*, and there is felt about the heart a certain oppressive *strictness* as if the orifices of it were drawn together, with a manifest *chilness* congealing the blood and communicating itself to the rest of the body.[30]

Such 'dejecting symptoms' had a long-term detrimental effect on the body; extreme sadness

darkneth the *Spirits*, and so dulles the *wit*, obscures the *judgement*, blunts the *memory*, and in a word beclouds the *Lucid* part of the Soul: it doth moreover incrassate the blood by refrigeration, and by that reason immoderately constringe the heart, cause the lamp of life to burn weakly and dimly, corrupt the *nutritive* juice and convert it into that Devil of a Humour, Melancholy.[31]

Then as now, different types of persons experienced greater or lesser degrees of emotion, though the propensity to experience those was

explained in physical or spiritual, rather than psychological, terms. For
not only the physical constitution of the body, its humoral composition, but also the soul's misapprehension or misjudgement could be
involved, resulting in extreme cases in mania, melancholia, or madness. In this context, emotional imbalances—like melancholia—were
both mental and physical diseases.

Mechanical Philosophy and Hydrodynamic Principles

The 'new physiology' of the seventeenth century—usually said to
have begun with William Harvey's discovery of blood circulation
in 1628—brought considerable changes to medico-scientific understandings of emotions, as to all aspects of human physiology and
psychology.[32] It was not unconnected that new ways of thinking
about the human frame—which separated out mind and body more
than had been previously conceptualized, and which prioritized the
former over the latter as the origin of knowledge and consciousness—coincided with a potential secularization of the body and the
emergence of rational philosophy as a way of understanding how
humans thought and felt. The influence of Isaac Newton in this
endeavour is well known. Newton's *Philosophiae Naturalis Principia
Mathematica* (*Mathematical Principles of Natural Philosophy*), published in
1687, laid out a new science of dynamics that would be applied to
many other fields, including science and medicine.[33]
   The appeal of Newton's method lay in providing a mathematical
approach that could be applied to rational mechanics. It gave rise to
various new schools of rational medicine, including iatromathematics,
iatromechanics, and iatrochemistry, as well as contributing to Cartesian principles and mechanical philosophy.[34] Indeed, the rationalist
physician George Cheyne, author of such influential works as *The
English Malady*, argued in 1701 that what medicine needed above all
else was a '*Principia Medicinae Theologia Mathematica* based on Newton's *Principia*—one that would integrate medicine and mathematics'
beyond all doubt.[35] Debates between rationalists (who developed systems of thought based on first principles and theories) and empiricists
(who derived their findings and theories from observation and analysis)
continued throughout the eighteenth century.[36]

Whilst theologians and moral philosophers discussed the asso-
ciative and cognitive dimensions of emotions, understandings of
their function and structure were influenced by medico-scientific
debates on the material structure of the body. This included the
iatrochemical approach of Thomas Willis; the mechanical physiology of Archibald Pitcairne, a physician and mathematician known
throughout Europe, and of Friedrich Hoffman; the hydro-dynamical
circulatory physiology of Boerhaave; the models of sensibility and
irritability found in the work of Von Haller; and the neurophysiology of William Cullen.[37] The combined impact of their research
helped to redefine relations between corporeal and incorporeal functions at the level of theory, and facilitated a concept of the body
that operated according to distinct laws of motion, with or without the presence of a soul. Feeling and movement (sensibility and
irritability) were instead postulated as the phenomena characteriz-
ing life.[38] These shifts necessarily influenced broader metaphysical
considerations about the nature of emotion and of the mind-soul
relation.

Mechanistic physiology was pioneered in France by the French
philosopher and experimentalist Pierre Gassendi; the mathematician
and theologian Marin Mersenne; and especially by René Descartes,
whose *Les Passions de l'âme* offered mechanistic interpretations of
human sensation according to laws of matter and motion.[39] Descartes's
work is often believed to have laid the foundations for modern
neuro-physiological psychology, and he certainly set out the mind
as a separate focus for study in ways that were hitherto unrealized.
Although Descartes reaffirmed traditional physiological explanations
of the passions by retaining the animal spirits of the nervous system
as a key factor, he introduced the pineal gland as the material site
for the interaction of the soul with the body, and gave mechanical
explanations in place of traditional humoral notions (see Fig. 2).[40] In
focusing on the doctrine of the reflex—which provided a rationale
for the machine-like functioning of the nervous system and the
body it controlled—Descartes applied mechanical philosophy to the
functioning of the human body in ways that potentially removed
the soul from its position as causal factor in bodily changes, and made
it no more than a passive observer.[41] In this model, the soul was
independent of, and separate from, the corporeal body, though both
were mutually influential through a process of interaction.

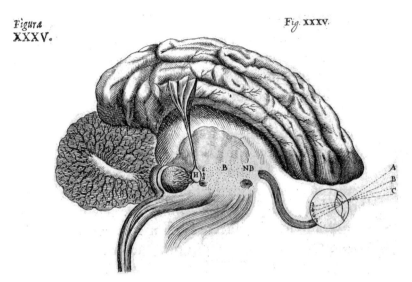

Fig. 2. Descartes and the nervous system.
Descartes: The Nervous System. Diagram of the Brain. Showing the location
of the pineal gland (H). From *De Homine* (1662).
Source: Wellcome Images.

By separating the 'principle of life', common to all animals, from
the 'principle of mind', by placing limitations on the mind's influence
on the body, and by presenting a distinction between mental and
bodily processes in a way that potentially divorced physiological and
psychological principles, Descartes presented a physical interpretation
of the origin of passions that was distinct from the approach of
Thomas Wright, discussed above. Consider the following synopsis
of Descartes's position, given by Susan James in her erudite work
on *Passion and Action*.[42] First, the animal spirits would move along
the nerves from the sense-organs to the brain. There, motions in
the cerebral cavities would push the spirits along the nerves, causing
such bodily events as a rush of blood to the area around the heart,
or the contraction of the muscles in the limbs. This instinctive
physical motion explains the behaviour of animals (such as sheep
that run from a wolf without feeling fear), for which passions are
unsophisticated reflex mechanisms because they lack souls. In humans,
by contrast, the motions of the cerebral cavity may move the pineal
gland (identified by Descartes as the site of interaction between

mind/soul and body), and thereby cause perceptions in the soul that may be sensory representations of passions.[43]

The same process was responsible for the physical effects associated with specific emotions, such as blushing in shame or paling in fear. Instead of being impulses of the mind/soul that were subsequently felt in the body, Descartes presented emotions as bodily happenings felt first in the *soma* and then presented to the *psyche* through the senses and processes of feeling and understanding located in the brain. In these terms they were 'passions of the mind'—as perceptions perceived by the mind and related to the mind 'caused, mediated and strengthened by movements of the nervous fluid'.[44]

Descartes's articulation of emotions as bodily experiences did not end debates over the role of the soul. Nor did it mean that the soul did not continue to be understood in many quarters as the originator of the passions and the mediator between *psyche* and *soma*. Following Boerhaave, some mechanists held the soul to be a rational, immaterial principle that was somehow attached to the material body, and that operated in accordance with established rules on matter and motion.[45] Vitalists and animists alike rejected the limited status this accorded to the soul, claiming that, rather than being an adjunct of the body (its physiological role restricted to 'willed activity and to consciously perceived sensations'), the soul or *anima* acted independently.[46] Indeed, the soul performed the ordinary functions of life in humans, although those of lower animals might be performed by mechanical principles.

As Geyer-Kordesch has observed, there was a unity of body and soul that challenged both 'somatically oriented medicine as well as a post-Cartesian philosophy'.[47] There was no separation of mind and body, or reason and emotion in the work of Stahl: emotion was connected to reason, and the imagination; the pre-Cartesian model of identity remained intact.[48] Later in the century, vitalism offered a more sophisticated approach, largely by replacing an abstract soul with the doctrine of sensibility, as seen in the work of the French physicians Paul Joseph Barthez and Théophile Bordeu. Whilst Barthez placed emphasis on a 'vital principle' that differed from the thinking mind and was the cause of life itself (a principle, moreover, that differed from the soul and was found not only in humans but also in animals and plants), Bordeu's contribution was to attribute glandular activity and other vital functions to 'sensibility', a vital force that resided in the material body and was not imposed on it from without.[49]

If Cartesian philosophy did not end debates over the role of the soul in emotional experience, it did provide the basis for alternative interpretations of human physiology to those found in humoralism. As such, it allowed a shift in physiological interpretation that was compatible with parallel developments in philosophy and physiology, including William Harvey's discovery of the circulation of the blood, the association principles of John Locke and others, and the development of nerve theory.[50] Descartes's work on emotions therefore needs to be understood in the context of medico-scientific change generally. In providing a way to think about emotions less in terms of humoral balance and more in accordance with the clockwork functioning of the body, mechanical philosophy popularized a language of ebb and flow in which new and modified versions of human physiology could flourish.

The late seventeenth century saw the emergence of hydro-dynamical theories of the body and mind. Scottish physicians such as Pitcairne and George Cheyne—Pitcairne's pupil who was equally well known for his attempts to practise what he preached in the field of medicine—conceived of the human body as a system of 'pulleys, springs and levers, pipes and vessels, its fluids being governed by the laws of hydraulics'.[51] For Pitcairne, the living body was composed of 'canals' and conveyed 'fluids' that moved with blood circulation: bodily disease could be viewed as deriving from an unusual circulation of blood in some or all of the parts of the body.[52] In this context it was possible to perceive of emotional extremes, such as melancholia, as the product of defective motions of the blood and an excessive accumulation in the brain, largely as a result of the affected flow and distribution of the animal spirits.[53] Again, the material construction of the body, rather than any psychological determinants, could be the originator of emotional experience.

Demonstrating the common practice of uniting older and newer theories of human physiology to formulate revised interpretative theories, Friedrich Hoffman used a notion of fermentation (characteristic of iatro-chemical discussions) in conjunction with a reliance on mechanical models. In his work the human body was conceived as a machine with solid and fluid parts. Although obstructions in the solids could interfere with the equilibrium of motion, it was the movement of the fluids and their constitution that determined health, disease, and emotional tendencies: 'in melancholics the spirits are indistinct and fixed, and approach a sort of acid nature. They not only leave enduring fixed ideas in the brain pores, but promptly uncover similar traces, of ideas of sadness, terror, fear, and so on.' When the body fluid became acidic and fixed, therefore, an individual subsequently became 'slow, timid, and sad'.[54]

Hoffman's influence in physiological theory increased over the several decades from the 1690s, an influence matched only by Hermann Boerhaave (whose work Hoffman did much to systematize) in the eighteenth century. Boerhaave developed a system by which most eighteenth-century medico-scientific research into emotions was conducted.[55] Indeed, the Edinburgh Medical School was established in 1726 along the lines of the system in place at Leyden, where Boerhaave was influential.[56] Boerhaave viewed forces as inherent in matter, expressed as mechanical movements and determined by mass, number, and weight. Forces were similarly expressed within the body by movement, contraction, and relaxation. According to hydro-dynamical principles of motion, Boerhaave constructed a system of physiology that viewed the body as a machine containing solid parts as its framework and circulating fluids that were charged by mechanical forces.[57] Health and disease were dependent on the circulation of the fluids around the body, as were extremes of emotional states. Hypochondriacal disease or melancholia, for instance, derived from the thickening of the fluids, and the accumulation of pathogenic material in the body from the bowels of the abdomen to the chest. As the vessels became obstructed and the brain received putrefied matter, 'the Vessels of the abdominal Bowels create a Stagnation, Alteration and Accumulation of black Choler which insensibly increaseth, though the Body was very healthful but a little before: And also that the same black Choler, when bred from bodily Causes, doth produce that Delirium.'[58]

The Role of the Nerves

Mechanistic and iatro-chemical and hydro-dynamical theories of the body moved away from humoral interpretations but continued to explain emotional experience in materialistic terms. Although the circulation of fluids round the body still held important meanings as under humoralism, eighteenth-century theorists reconceptualized the body in which the fluids flowed. It was the solids (nerves and fibres) rather than its fluids that were conceived as the 'true basis of the body'.[59] And, rather than being related to the apprehension of the soul *as* mind, emotions were redefined as a product of sensory perception and material processes. Nervous physiology allowed the main office of the brain and nerves to be 'sensation; that is, to suffer changes from the impressions of external substances in the parts of the body affected by them, and to undergo analogous changes in the representations of the mind'.[60]

There is an extensive literature on nerve theory during the eighteenth century, which I do not wish to revisit here.[61] What is relevant is that theories of the nerves began to explain the sympathetic response between heart and body as well as emotional susceptibilities and disorders. As William Clark, MD, Intra-Licentiate of the College of Physicians in London, put it in 1752, 'every Faculty of the Mind depends on the nervous system', and 'all the Functions of the Body, rightly exerted, depend upon a due distribution of the nervous Fluid'.[62] In this context, the physiology of fear, for instance, was attributable to an increase in nervous fluid, which contracted the muscles and signalled alterations in the motion of the blood and the heart. The exertion of the muscles in anger, or the 'palpitation of the heart in terror', similarly derived from 'exertion of the nervous power', the strength of which depended on the state of the nerves.[63]

We have seen that, under Cartesian dualism, the heart could be conceived as a mechanical organ, rather than as an active and intelligent entity that summoned choleric blood in anger and melancholic blood in sadness. The heartbeat had become a mechanical impulse, and yet it remained something more. It was in many ways the most measurable and noticeable of the vital actions, and yet it was simultaneously outside the control of the conscious and rational soul. It continued to beat outside the body, and so was removed from the soul in the brain, and yet it reacted to emotional experiences (that *were* regulated by the soul), beating excessively in love and anger; stuttering in fear.

These complexities concerning the heart were increasingly addressed in terms of the body-as-machine model of investigation. For Descartes, nervous spirits originated in the brain and were the agents of sense and motion. Though this was not very different, in essence, from traditional ideas about the spirits, its inherent separation of mind and body, and of heart and soul (the soul now being allocated a material site in the pineal gland), meant that the mind was identified as the location of emotional cognition. To recap: emotions might be felt in the body, but they were realized through the cognitive principles of the brain, and produced a bodily response, not by the summoning of blood around the heart (as in Galen), but in the movement of the nervous spirits that resided in the brain.

Understanding the role of the heartbeat in this new physiology was part of the originality of Boerhaave's work. Whereas the heart was in essence the motor for circulation and the blood's movement under mechanical theory, that did not explain the origin of movement (the principle point of contention for the vitalists and mechanists). Boerhaave argued that nervous spirits were transmitted from the brain to the heart through a series of nerves that passed between the ventricles before entering the muscular heart. There the nerves were compressed through contraction, and the spirit's flow halted, before the cardiac muscles were relaxed, and the spirit resumed its flow. As Frank has argued, this description could not account for the vagaries of the heartbeat during illness or even emotional exertion, and so its acceptance was limited. More successful in explaining these problems was Von Haller, who replaced it with a form of the stimulus and response theory that depended on the varying degrees of stimulation inherent in blood.[64]

Nervous transmission provided a replacement for the movement of the spirits around the body and between the body and mind, and a secular language in which emotions could be expressed, redefined, as they were, as products of sensory perception and material processes.

Concepts of nervous transmission also offered theorists and experimenters a way to understand the mind-body relationship that took account of such problematic phenomena as emotions, and provided an explanatory structure for theories of 'mind'. This was no instant 'fix' to the problems that had been left by the erosion of humoralism: such concerns as the precise location of the soul, and the relation between the soul, the nervous system, and the mind remained contentious. Yet the nervous system provided the 'interface' between the otherwise disjointed physical and psychical realms.[65] It also helped to explain diseases of the mind and the body: since the quality of the solids and the circulation of the fluids dictated mental and bodily health, obstructions in either produced illness, because of interdependence between connecting parts. For this reason, the physician and natural philosopher Robert Whytt famously claimed that 'all diseases may, in some sense, be called affections of the nervous system, because in almost every disease the nerves are more or less hurt; and, in consequence of this, various sensations, motions and changes, are produced in the body'.[66]

These developments in the field of medico-scientific theory linked to developments in moral philosophy and early psychology to make it possible for emotions to be viewed as primarily physical phenomena. The external senses—sight, smell, hearing, taste, touch—transmitted ideas of 'external substances' to the mind, the 'nervous power' instigated to respond being responsible for the effects of the passions:

which, if lively and exhilarating, relax the influx of blood; and by the remission of the nerves; hence redness, moisture and turgescence of the skin. Those passions, which are languid and depressing, contract the exhaling vessels; as appears from the dryness of the skin, produced by them: from the goose-skin, by terror; and from diarrhoea, caused by fear.[67]

The ability of emotions to cause physical illness was explicable because 'they also seem to dilate the inhaling vessels, whence fear facilitates the action of the smallpox and the plague'.[68] Whereas the external senses 'being affected by external objects, [transmit] some change by the nervous spirits' to the brain, the internal senses—including thought, imagination, and memories—were similarly responsive to external qualities (being 'impressed in the body itself, and indeed in the medulla of the brain') by objects of sight, smell, hearing, taste, or touch.[69] A lack of judgement or irrational passion could be provoked by the lack of a 'healthy constitution of the brain. For when that is compressed, irritated, exhausted of blood, or changed in its fabric,' the faculty of judgement failed.[70]

It would be naive to suggest any wholesale transition of emotion theory from mental to corporeal causation during the long eighteenth century, or back again during the nineteenth. In classical medicine it is possible to find physicians ascribing emotional states to physical causes—as in Hippocratic assertions that the heart might become filled with blood because of impeded menstruation, thus causing stagnation and putrefaction of the blood, suicidal thoughts, fear, and distress.[71] Moreover, under nervous physiology there remained the possibility of some passions being linked to mental apprehensions, some essence of understanding that was distinct from material processes. Von Haller, therefore, allowed that there were 'affections of the human mind' that were linked to some innate perception of what was beneficial or detrimental to an individual's safety or health: 'the presence of good constitutes joy; the desire of good, love; the expectation of good, hope; the presence of evil, sorrow, terror or despair; the dislike of evil, hatred; and the expectation of evil, fear', all of which were somehow 'unconnected with the properties of matter, or certainly less simple, understood or mechanical' than other passions.[72] Other medics, including Dr Corp, MD of Bath, abstractedly, and without explanation, wrote of certain emotions—'Hope, Joy, Anger, Fear, Grief, Anxiety'—as '*purely mental*, as originating in the Mind, and not excited by, or blended with, corporeal sensation'.[73]

Although there continued to be a school of thought in eighteenth-century medical culture that viewed emotion as primarily mental phenomena with bodily effects (the 'common-sense' model with which we are most familiar), most viewed emotions as material entities, produced by the condition and structure of the *soma*. Their emphasis on the body's constitution meant that an individual's susceptibility to degrees of emotion differed according to their physical predisposition, mental apprehension being dependent on the physical sensations. To this end the influential physician Thomas Cogan published a five-volume *Treatise on the Passions and Affections of the Mind* (1813), which asserted that passions derived from some 'originating cause' (perception or idea) that 'violently agitates the corporeal frame' and causes a 'change in the state and disposition of the mind'.[74]

The belief that emotional experience was linked to corporeal characteristics led to a view of emotional types that were identifiable not in terms of an individual's humoral propensity, but according to the state of his or her nerves and fibres. Those with a 'large brain and thick strong nerves' possessed 'a great sensibility as well of the whole body as of the organs of sense. Hence arises a ready apprehension of objects, and an increase of understanding and knowledge ... which choleric persons possess in so eminent a degree; but along with this condition of the nerves, they are excessively liable to grief and anger.' By contrast, according to Von Haller, those with a 'small brain and slender nerves' possessed 'senses more dull and a phlegmatico-melancholic torpor conjoined'.[75]

Emotional predisposition could be impacted by environmental factors that modified the condition of the nervous system. The condition of one's emotional predisposition, or 'temperament', to use a controversial term that originated in the seventeenth century, but came into increased usage by the eighteenth, influenced the condition of one's heart, just as it had under humoralism or mechanism.[76] An examination of the case of John Hunter, whose fatal attack of angina pectoris was linked to his tendency to experience anger, provides a case in point.[77]

Much has been written about the class-based diagnoses of hypochondria as an affliction, and the influence of habits of living—emphasizing the 'non-naturals'—was stressed in literature on emotion. Thus 'persons of gross full habits, the robust, the luxurious, the drunken, and they who sup late' were most likely to succumb to excess fears and even nightmares.[78] Of course, the gendering of hypochondria was also significant. Like those of children, the fibres of women were generally 'lax and soft, the nerves extremely irritable, and the fluids thin'.[79] Whilst old age brought a rigidity of the fibres for both sexes and an insensibility of the nerves—giving rise to very different diseases and treatments from those that attacked the young—female constitutions, ordinarily prone to as 'peculiar sensibility of the nervous system', were regularly disturbed by menstruation.[80] Nervous debility and emotional complaints frequently plagued 'women who are obstructed; Girls of full, lax habits, before the eruption of the Menses'.[81] This was especially the case during menopause, 'when Women pass the fruitful seasons of life, and the delicate uterine Tubes contracting themselves, become too rigid, and resist the impetus of the Fluids so as to prevent the usual discharges'.[82] Under nerve theory, then, the gendering of emotions seemed more fixed than ever before.

Electrical Transmission and the Neuro-Sciences

Contemporaneous with the growth of nerve theory from the eighteenth century were debates on animal electricity, and over the existence or otherwise of a 'neuro-electrical' fluid that flowed in the nerves and caused contraction of the muscles.[83] Since the experiments conducted by Luigi Galvani in the 1790s, it was difficult for theorists to shake off the idea that the blood possessed an animating principle, and that the muscles could contract after the heart stopped beating. Discussions of electrical communication throughout the body, and between the body and the mind, also facilitated the rise of mesmerism in Victorian Britain.

Alison Winter has discussed the popularity of mesmerism at all levels of society from the early nineteenth century. The creation of the eighteenth-century physician Franz Anton Mesmer, who applied theories of 'animal magnetism' to healing and health practices, mesmerism demonstrated the interconnections between mind and body and, somewhat controversially, the ability of psychic energy to heal the *soma*.[84] While the mind sciences and physiologists began to attend to the meanings of the brain and emotional responses as matters of science, advocates of electrical medicine 'began to see the body as a battery, storing and dispensing electrical influence as needed; the new "reflex" physiology represented the human body as a system of switches and levers, reflecting incoming stimuli outward again in bodily action'.[85] In the process, Winter argues, distinctions of mind and body became (perhaps temporarily) irrelevant: the field of 'mental physiology' explored links between the two areas of mental and physical phenomena.[86] Interpretations of the mind–body relation were at the centre of the mesmeric enterprise, just as they were at the centre of debates on the emotions and the heart.

Of Mind and Matter

The mind–body relation has long been a subject of interest to Western philosophers and physicians.[87] Conceptualizing consciousness (however defined), and the interactions between thought, 'feeling', and matter (and action) has a historiography that dates back at least as far as the ancient Greeks. By the seventeenth century, there were many literary, scientific, and mystical texts that debated the respective qualities and responsibilities of the mind and body; many of these focused on the mind's ability to heal (or to harm) the body.[88] In these texts, 'mind' was defined as a 'spiritual essence' akin to the soul, and the physiological processes associated with certain emotions were processes that helped maintain a natural and—for many seventeenth-century writers—a chemical equilibrium.[89]

Of crying or weeping, for instance, the Flemish chemist, physiologist, and physician Jean-Baptiste Van Helmont observed that 'persons overtaken with some great grief or affliction, when they cannot discharge their Sorrow by weeping, do often fall into some Distemper or Sickness, because the Idea of the Cause of their Sorrow by this means encreaseth'.[90] Similarly, the heart and the mind and body were seen to work together in concert with the will:

Notwithstanding, that Man's Body consists of many different Members, which disagree in Figure and Operation; yet we find that all the said Members do co-operate together in Concord and Harmony. Whence we may conclude, that all the Spirits that are within the same Body, are governed by one central Spirit . . . As by Example, when I have a mind to go, speak, or look about me, I find that my Will, which proceeds from my Heart, and is wrought out in my Head, causeth the Spirits that are in the Members, to put them into such a Motion, as is answerable to the Will and Command of my Principal or Ruling Spirit.[91]

As is evident from this extract, the heart formed a crucial role in mediating between the mind and body as separate but interrelated areas, and the mind itself was a spiritual essence, rather than a material mass of neurons (as it would be conceived of by the twentieth century). The processes by which that change in interpretation occurred—a change most commonly associated with the French philosopher René Descartes and the popularization of Cartesian dualism—is central to the history of the emotional heart of medical and cultural theory.

It had been argued that debates over the relationship peaked during the period of the Enlightenment, when both mind and body received an unprecedented degree of attention from both philosophers and physicians alike.[92] Despite attempts by many to rationalize and make scientific the organ of the heart from the late eighteenth century—by making it an organic structure prone to decay, by divesting it of its moral or ethical leanings, and by emphasizing its significance to the material circulatory and nervous systems, as detailed in the following chapters—there was one small and persistent problem: the heart's vitality. Traditionally, it had mediated between mind and body, between experience, action, and will, as seen in the above extract from Van Helmont's *The Spirit of Diseases*. The final part of this chapter, then, will examine some of the philosophies of the mind-body relation in which the heart has been accommodated since humoralism. As we will see, the heart remained an important source of mental and physical influence well into the nineteenth century, with one important difference: an integral part of the search for mind-body understanding became the quest to replace the soul.

Connections and Conduits

As outlined above, the humoral tradition provided a physical causal explanation for emotional extremes that was rooted in an imbalance of the body's fluids. Excess humours could impact on the emotions and vice versa. A classic example is hypochondria, a disease that was protean in nature, and took a variety of forms from constipation to heart palpitations to fits and depression.[93] With historical and symbolic relevance stemming from antiquity, the heart was central to these discourses on emotional embodiment and pathology. As such, it partook of the energy and motive power of the soul as it moved in and through the physical body.

In addition to humoral influence, there were other conduits of mind-body interchange available to eighteenth-century medical commentators. One of these was a localization theory: that individual organs of the body, such as the uterus and the digestive tract, could disorder the brain. These vague hypotheses would be concretized in the latter part of the century by the work of the Paris anatomist Marie François Xavier Bichat, who referred to the ganglia of the autonomic nervous system, the gastrointestinal tract, and the brain as intricate pathways for nervous disease that affected the body and the mind.[94]

The idea of nervous disorder and nervous disease impacting negatively on the brain in particular was of lengthy duration. It was popularized in the late seventeenth century by the German physician and chemist Freidrich Hoffmann, whose work on the body (discussed above) hypothesized a 'nervous ether' that radiated out from the brain and that set the rest of the body mechanically in motion.[95] By the time that Cullen added neuroses, or diseases of the nervous system, to his popularized nosology of disease classes, the principles of nervous disease were relatively widely disseminated.[96]

Concepts of neuroses did not have the same meaning for Cullen and eighteenth- and nineteenth-century medical practitioners that they would have for Sigmund Freud and twentieth-century mind scientists. For Cullen, 'neurosis' applied to any disease or disorder, including cardiac neuroses, that was characterized by abnormal function that was rooted in the nervous system. The ability to locate psychological disorders in the structure of the body would provide a material explanation for mind–body dualism. By the following century, 'neuroses' would begin to describe mental, rather than physiologically influenced, conditions; by the time of Freud, they had become the 'outward garments of repressed desires'.[97] This almost complete reversal of the meaning of neurosis as applied to mental, rather than physical, causation, was one way of reinterpreting the mind-body relationship that was realized in the nineteenth and twentieth centuries.

By the 1870s the ubiquity of nervous disorders that identified a range of functional and emotional disorders—and the popularization of the American neurologist George Miller Beard's concept of neurasthenia—meant that, at the same time as emotions were regarded as mental phenomena, they were also, paradoxically, more linked to the body than ever before through its originating principle and a complex symptomology.[98] That unreliable nerves—and emotions—were increasingly linked to women, through the doctrine of neurasthenia, is evidence of the gendering of emotions and of nervous diseases throughout the nineteenth century.[99]

Philosophies of Difference

Accompanying the physical separation of mind and body that was discernible in medical and scientific theorizing were a range of philosophical interpretations. At the same time as physicians treated the mind and body holistically—that is, recognizing the mutual influence of the psychic and the somatic in human health—some writers attempted to demarcate and make explicit the boundaries between them.[100] These included John Petvin's *Letters Concerning Mind* (1750), John Richardson's *Thoughts upon Thinking* (1755), and John Rotherham's *On the Distinction between the Soul and the Body* (1760).[101] The idea that the body was separate from the mind, then—and, equally significantly, that the mind was separate from the soul—was a matter of intense philosophical speculation from the mid-eighteenth century. In the following century this trend would become more elaborate, with the measurement of mind principally being involved with a secularized and increasingly anatomized brain.[102]

Of course, the impulse of change was, and is, seldom unidirection-al. Nor does it take place without resistance or counter-models.[103] There were many alternative models to Cartesian dualism from the eighteenth century. There were also ways by which mind-body dual-ism was given a familiar linguistic and philosophical footing (largely through the use and adaptation of concepts of neurosis, which, as discussed above, provided first a material and then a psychologi-cal explanation for many problematic phenomena). One interesting example of an alternative philosophical model to the heart of mate-rialism was the rise of romanticism in the literary and artistic sphere, and of psychosomatic influence in the field of medicine.[104]

While romanticism arguably still holds sway over cultural attitudes towards affective experiences such as romantic love, self-hood, and emotional expression, psychosomatic models have continued to influ-ence attitudes to heart health well into the twentieth century.[105] As Rhodri Hayward has recently argued, it is commonplace for modern medicine to regard the body as a 'moral touchstone, in which the sins or troubles of the past come out in the flesh'.[106] As this chapter has shown, this is not a new phenomenon. The liberation of the mind from the body in medical theory took place over a relatively short historical period, and was essentially a product of nineteenth-century (particularly *late* nineteenth-century) theorizing.

Moreover, historical attempts to disengage the mind from the body have often been disrupted by religious and spiritual principles. Edward Stainbrook notes that, although the nineteenth century saw the liberation of medicine from many traditional theological and moral restrictions, theories of the mind were still 'largely bound to religion and to philosophy'.[107] Moreover, the conception of mind that emerged from the nineteenth century—represented by a shift from the study of mind and soul towards the brain and the nervous system—continued to support a psychophysiology of mind-body functioning that made material the links between mental and bodily processes. In most cases, though, the organ of feeling had become the brain, rather than the heart. The heart responded but did not produce emotional experiences; in this it was the most 'sympathetic' organ of the body. While one might say 'I love you with all of my heart', the physician Daniel Hack Tuke explained, that was not because 'those sentiments are produced in the heart, but because in every violent affection, either the heart or other parts, by the movement by which we describe the affections, in our language, *act sympathetically*'. This opposes the 'vulgar error that the heart is the seat of our passions'.[108] By 'sympathetically', of course, Tuke was referring to a complex physiological response that was embedded in the body's tissues and fibres, as opposed to modern sentiments connected with pity or emotional sensitivity.[109]

From the Nervous Heart to the Heart's 'Little Brain'

Since the twentieth century, discussions of the mind-body relation have continued to move away from cardio-centric and towards neuro-centric understandings of emotional experiences, at least in the West. And yet there have been several interesting attempts to resolve mind–body dualism. The emergence of endocrinolological-cal explanation, for instance—focusing on the nature and effect of hormones—provided a materialistic account of the mind-body relationship in ways that echo the humoral interpretations of the ancients.[110] In 1905 the English physiologist Ernst Starling first referred to a substance he would later describe as 'hormones' (from the Greek, and meaning 'to arouse' or 'to excite') as 'chemical messengers' that were circulated around the body to communicate between its various parts.[111] The subsequent and widespread acceptance of the hormones as an explanatory category seemed to elucidate many of the communications that took place between the brain, the heart, and the rest of the autonomic nervous system.

Parallel and divergent accounts that addressed the mind-body relation included psychosomatic theorizing (particularly in relation to psychoanalysis) and a new 'holist-mentalist paradigm' demonstrated by neuroscientific investigations into consciousness.[112] Notably, and despite their divergent positions, each places an emphasis on the mind (located in the brain) as the repository of emotions—embodied emotions are expressions, rather than causal agents, of affect. The heart, moreover, has been notably absent from analysis.

However, the relationship between the brain and the heart has continued to be an issue in modern medical theorizing. Such concerns as the origin of life and the mechanism of the heartbeat are inevitably wrapped up in physiological investigations into the structural connections between the brain and the heart, and the nature of cardiac tissue. In recent decades, for instance, organizations like the *HeartMath Institute* have used scientific language and reasoning to argue for the existence of a 'little brain in the heart'; the heart, it is argued, is not a mere lump of muscle, but a bundle of neurons and ganglions—similar to those in the brain—through which this organ communicates with the brain and influences information processing, perceptions, emotions, and health.[113]

The answers to many of our original questions now provide a scientific basis to explain how and why the heart affects mental clarity, creativity, emotional balance and personal effectiveness. Our research and that of others indicate that the heart is far more than a simple pump. The heart is, in fact, a highly complex, self-organized information processing center with its own functional 'brain' that communicates with and influences the cranial brain via the nervous system, hormonal system and other pathways. These influences profoundly affect brain function and most of the body's major organs, and ultimately determine the quality of life.[114]

HeartMath might use a scientific formulation for evaluating the impor-
tance of the heart in the mind–body relations; Eastern philosophies use more 'traditional' holistic strategies, as seen in Ayurvedic and Chinese medicines.[115] What is interesting from the perspective of heart history is that these interpretations and principles of therapeutic practice are gradually making their way into the Western orthodox tradition.[116] It is interesting to observe that the mind–body holism of the Galenic world is having something of a revival.[117]

    The heart's redeployment as an organ of emotion, whether expressed in the language of scientific materialism, or as an eth-ereal or spiritual 'presence', must indicate something of a backlash in Western scientific medicine against medical specialization and the heart's identification as a material pump, or muscle. It also indicates the ideological, philosophical, and theological weight of resistance *against* viewing the body and mind as separate or divisible worlds. These themes emerged again and again in eighteenth- and nineteenth-century medical practices, with scientific medicine being juxtaposed against traditional beliefs about the heart's function. In the following chapter, that struggle is played out in the body of the most influen-tial pathological anatomist and surgeon in eighteenth-century Britain: John Hunter.

2

Hunter's Heart: Pathological Anatomy and the Science of Disease

The stand of coaches in the Palace-yard intercepted his passage, and he bid one of the coachmen to make way for him. The fellow refused, and became insolent, and John Hunter losing all temper, gave vent to the most terrible execrations, which only produced laughter in the other... When he arrived he sat himself down, saying, the rascals have killed me, and Mr Heaviside supported him in his arms, expecting every moment to see the first anatomist in the world expire in this untoward situation.

Robert John Thornton, 'Life of John Hunter', in *The Philosophy of Medicine* (1799–1800)[1]

The Heart of Hunter

On 16 October 1793, the surgeon and anatomist John Hunter attended a board meeting at St George's Hospital in London. The meeting took place after a series of quarrels and confrontations between Hunter and the other board members. When the meeting grew confrontational and the discussion heated, Hunter found his opinions rebuffed. According to onlookers, he 'immediately ceased speaking' and left the room. Apparently struggling to 'suppress the tumult of his passion', he had scarcely reached the privacy of an adjoining room when, 'with a deep groan, he fell lifeless into the arms' of a colleague.[2]

An autopsy was performed on the body of John Hunter by his brother-in-law, Everard Home. Home identified the cause of death as a diseased heart, a result of angina pectoris: the carotid and coronary arteries and their branches being 'thickened and ossified'; the heart 'the chief seat of disease'. The pericardium was unusually 'thick', though 'the heart itself was small, appearing too small for the cavity in which it was contained, its diminished size being the result of wasting'.[3] Home's conclusion was that Hunter's heart was 'unable to carry out its functions, whenever the actions were disturbed', either 'in consequence of bodily exertion' or of 'affections of the mind'.[4] The most recent spasm stopped the heart, pressing the nerves against the ossified arteries, and preventing it from resuming its work until it was too late: 'Death immediately ensued.'[5]

The sudden and fatal collapse of John Hunter, and his death from heart disease, have provided an important narrative framework for understanding certain aspects of his life and career. Like his brother William, John Hunter has been the subject of several biographical studies, and their individual achievements and rivalries have received considerable historical attention. By the time of his death, John Hunter was the most famous surgeon in eighteenth-century London. He was also a successful naturalist, anatomist, and collector of wet and dry specimens, many of which are now housed at the Royal College of Surgeons' Hunterian Museum (Fig. 3).[6] He had published extensively on his research and, despite his early lack of professional training and educational difficulties, gained considerable status in his field. As with his brother William, historians have assured John's reputation: he has been credited as the originator of the modern sciences, as the unifier of pathology, physiology, and therapeutics, as the founder of modern biology, as well as the 'father of modern surgery'.[7]

Discussions of the character and personal life of John Hunter have necessarily been influenced by historical context; as Stephen Jacyna has observed, contemporary constructions of Hunter as a man of science derived from a particular polemical context.[8] Part of his construction as a 'man of science', however, is also based on modern assumptions about what success requires. Most biographers have identified Hunter's obsessive, almost pathological, tendency towards hard work: 'commencing his labours in the dissecting room generally before six in the morning', and staying there until nine, when he breakfasted. He then saw patients, before returning to work well into the early hours.[9] Wendy Moore has recently, like many, implied that Hunter's 'drive' was because of his struggles with his brother, a narrative of sibling rivalry so familiar to modern readers.[10]

Fig. 3. The Hunterian Crystal Gallery today.
Source: The Hunterian Museum.

As will be shown in this chapter, the attribution of these psychological attributes made a significant contribution to the mythology
around Hunter's ill health and death, while his tendency towards
'irascibility' was attached to some gendered concept of the construction of a 'man of science'. It is no coincidence that contemporaries
endeavoured to 'account for the irritability by which men of genius
have so frequently been distinguished. They are, for the most part,'
Dr V. Knox concluded, 'in a state of intense thought [and] . . . every
little accident is likely to disturb the repose of him who is constantly
engaged in meditation, as the string which is always kept in a state of
tension, will vibrate upon the slightest impulse'.[11] It is worth noting
that the imagery used here to describe physiological process—through
strings that might vibrate—relies heavily on the implicit assumptions
of nerve theory as a way of understanding the connections between
mental and physical processes.[12]

These kinds of assumptions about Hunter's personality, then, the
story of his success as a man of science and that of his chronic ill
health (he had suffered periodically from cardiac symptoms in the
years leading up to his sudden death), tell us much about the meanings
of the heart and its links with emotion in the eighteenth century.
It is no coincidence that angina pectoris had been identified only a
few short years earlier, and Hunter was one of the men responsible
for identifying and naming the disease. When a 54-year-old angina-
pectoris sufferer died suddenly in 1775 and—like Hunter—'in a
sudden and violent transport of anger', his physician, John Fothergill,
had requested that John Hunter 'open' the body.[13]

In his dual capacities as morbid anatomist and surgeon, Hunter
participated in the process by which the body and its physiology
were taken apart from the eighteenth century, its functions and
dysfunctions identified, named, and classified. His research revealed
discoveries in a variety of areas that included placental function,
shock, blood coagulation, the treatment of wounds, venereal disease,
tissue transplantation, and valvular heart disease. One of the most
important concerns behind his catalogue of work, and his collection
of comparative anatomy specimens, was to establish rules of anatomical
similarity, the relationship between structure and function, and the
formulation of general laws of physiology and pathology. These areas
were explored through analysis of the nervous system and brain
anatomy, and the anatomy of the digestive system and the heart.[14] It

is perhaps ironic, then, that Hunter's own heart has received so little historical attention. We are accustomed to understanding emotional conflict as registering in the heart, or as having an impact on the heart's function (this is particularly relevant to modern-day concepts of 'stress'), and yet its meaning for the death of John Hunter has apparently been neglected.

In a major theoretical transition from the earlier period, the eighteenth century saw the heart of science isolated, disembodied, catalogued, and dissected, its material qualities and characteristics weighed and assessed as part of a complex and (to varying degrees) mechanistic model of the human body.[15] Unlike the heart of the seventeenth century (viewed from afar and in the context of well-established conventions around the mind, the body, and the soul), the heart of the eighteenth century was, above all, a material structure. And yet, particularly at the level of lived experience and medical therapeutics, it continued to be invested with emotional meanings.

Although the language of influence had changed—first, mechanistic forces and chemical influence, then the doctrine of nerves and sympathy, providing the tools of analysis rather than the fluids and balances of humoralism—emotions remained primarily bodily events. There was not yet any indicator of their shift to the mind, nor of the mind being secularized, as would take place in the rise of the nineteenth-century mind sciences. Though no longer wedded to the heart, they caused disruption to the organ when excessive, but in the same way as other extremes, mental, physical, and environmental. As such they could be managed only by the careful maintenance of the non-naturals and by the retention of holistic principles that were incompatible with Hunter's immoderate lifestyle. Before we can examine these claims in any detail, however, we need to return to the disease that killed John Hunter, and to the emotional meanings of angina pectoris.

Identifying Angina Pectoris

When the London physician William Heberden identified angina pectoris as a specific disease in 1768, it was in a paper given to the Royal College of Physicians and subsequently published in the Society's *Transactions* of 1772. John Hunter had conducted a largely inconclusive autopsy. The term 'angina' (derived from the Latin and meaning spasmodic and choking or suffocating pain) was not itself original, but, when added to 'pectoris' (meaning of the chest), it became a complaint.[16] It also denoted the first in a series of categories of cardiac diseases to emerge from the late eighteenth century.[17] Angina pectoris was, said Heberden, 'marked with strong and peculiar symptoms, considerable for the kind of danger belonging to it, and not extremely rare'.[18]

A series of casebook entries and examples testified to the ubiquitous and regular nature of angina–pectoris symptoms, which included severe chest pain, typically running down one arm, and a 'sense of strangling, and anxiety'.[19] The sufferer would typically be struck when walking. He (and it was usually a 'he') was often 'surprised with a fixed pain at the breast, which gradually, as he proceeds, increases, till at last he is obliged to stop lest he should die'. William Butter's 1791 *Treatise* continued:

On standing still this symptom abates, and is entirely gone in a few minutes, especially if he belch wind. This solitary pain is compared to a cramp. Most frequently, however, the fixed pain at the breast not only extends to other parts but is also preceded and accompanied by other symptoms. It is sometimes called an aching, sometimes a smarting, and sometimes a sharp pungent pain. Some call it a violent pain that cannot be described; others call it a numb pain; and others, a numb pain accompanied with a sense either of heat or cold indiscriminately.[20]

It was generally believed that there was something *specific* about the pain of angina pectoris, some quality that marked it out from other diseases. There was 'something which is beyond the nature of pain, a sense of dying', as the cardiac physician Peter Mere Latham put it.[21] *The Cyclopaedia of Practical Medicine* (1833), declared that there was 'something peculiar in the pain' of angina pectoris, 'whatever be its degree, unlike the pain of other parts of the body, and as if it were combined with something of a *mental* quality. There is a feeling and fear of impending death'.[22]

The most influential treatise to be published on angina pectoris after Heberden's initial description was Caleb Hillier Parry's *Inquiry into the Symptoms and Causes of the Syncope Anginosa, Commonly Called Angina Pectoris* (1799), a paper that was first read at a Gloucestershire Medical Society meeting in 1788.[23] That meeting was composed primarily of friends and colleagues of Parry, and included Hunter's friend and contemporary Edward Jenner. The *Inquiry* included case studies (some of which were presented by other physicians), which recorded the symptoms and deaths of patients who had been afflicted by this peculiar cardiac complaint. It provided a detailed and systematic account of diseases of the heart that could be linked to angina pectoris (associated by Parry with 'syncope' or loss of consciousness), and provided an extensive assessment of the findings of other physicians. Some of the causes given were structural—that is, they were based on pre-existing material conditions within the heart as an organ. These included tumours, ossification (the natural process of bone formation), and too much fluid in the pericardium.[24] There were also other 'Accidental, Occasional, or Exciting causes', and these were widespread. Though they included the passions or emotions, there were many other factors to consider, of which:

The chief are, certain circumstances of sensation, including the existence, and even the sudden cessation of bodily pain; the emotions of grief, joy, fear, disgust, and sympathy, more especially when suddenly excited; affections of various other parts of the body, particularly the alimentary canal; exposure to great external heat; different degrees of bodily exercise; the action of kneeling; the rising into an erect posture, after long confinement in bed by disease; the sudden removal of the fluid in the ascites, and of the foetus in delivery; want of food; sudden or great evacuations of blood; violent evacuation by stool.[25]

Explanations of the cause of angina pectoris clearly varied widely, and this variation was a phenomenon that became more important in the following century, with a greater degree of classification than we find in Parry's analysis. In the main, however, commentators believed that it was aggravated most commonly by extremes of exercise and emotion. One of the defining characteristics of the disease was rapidly understood as a tendency towards anxiety and anger, which placed a mechanical or hydraulic pressure on the circulation of the blood around the body. Angina attacks therefore often followed on the arousal of the body through exercise or emotional excesses, or any other condition that forced the blood to move rapidly around the body, and delivered too great a load on the heart. This was particularly problematic when there were pre-existing structural problems within the heart itself: calcified or corroded arteries that fed it, perhaps, or a flabbiness of the muscular tissue.

Of all these causes, emotions were the most common and consistently raised factor, perhaps because, while it was easy enough to avoid exercise, anger or anxiety were harder to combat. It is interesting, therefore, that Hunter's fatal attack developed immediately after his anger at a board meeting, for this provides an important narrative framework for understanding attitudes towards angina pectoris as an emotional disease. The circumstances surrounding the death of John Hunter have been used, perhaps anachronistically, given the relative 'newness' of stress concepts, as evidence of the intensity of passion with which Hunter approached his surgical and anatomical endeavours, and the competing tendencies of surgeons and physicians during these formative years of professionalism.[26]

Many contemporary and historical accounts of John Hunter have subsequently focused on his difficult and confrontational personality, and his tendency to quarrel with colleagues. In a discussion posthumously printed in the *Lancet* in 1839, Thomas J. Pettigrew reported that John Hunter 'was remarkable for his irascibility through life and it probably served to shorten the duration of his existence'.[27] In a section detailed 'Temper and Personal Appearance', he reported that Hunter 'had no command over his Temper'; that 'his speech was rude, and he habituated himself to the disgusting practice of swearing'. Some physiognomic link was perhaps alluded to between such brutish demeanour and his physical presence: he was described as average size, 'vigorous and robust', but with a short neck and 'rather large features'. His hair was 'reddish' in his youth (perhaps a reminder of the traditional association between red hair and hot temper) but became white in later years.[28]

Hunter's long-time rival Jesse Foot accused him often of 'exciting jealousies and quarrels amongst his colleagues', stating that he was embroiled in 'continual war' at St George's. This statement has been accorded little credit, given the antagonistic nature of their relationship.[29] Yet there is, elsewhere, ample evidence of what was perceived as Hunter's irascible temperament. Wendy Moore has given several examples of Hunter's disagreements with family members, employees, and colleagues.[30] His friend Lord Holland, for instance, reported that Hunter's judgement was clouded by an 'irascible and tenacious temper', and that he tended to be 'dogmatic and angry' when crossed.[31] And in his work on *The Philosophy of Medicine*, the physician Robert John Thornton reported Hunter's visit with John Heaviside (fellow surgeon and collector), to hear a trial at Westminster Hall. His observations, reported in the epigraph to this chapter, confirmed that Hunter was prone to 'the most terrible execrations' with the least provocation.

An important construction of Hunter's death from angina pectoris was clearly his inability to keep his temper, his tendency to become angry in the face of little provocation. Hunter himself famously declared that his life 'was in the hands of any rascal who chose to annoy and tease him'.[32] He expressed more than once the tendency for his 'spasms' of pain to come on during moments of anxiety in particular:

> The spasm on my vital parts was very likely to be brought on by a state of mind anxious about any event . . . I have bees . . . and I once was anxious about their swarming lest it should not happen before I set off for town; this brought it on . . . I saw a large cat . . . and was going into the house for a gun when I became anxious lest she should get away . . . this likewise brought on the spasm.[33]

The type of emotion experienced mattered, as it had in humoral accounts, with 'negative' feelings like anger and anxiety having a particularly detrimental impact on the human body. As legend has it, moreover, Hunter predicted his death at the board meeting, having expressed that very morning to a friend his apprehension in 'undertaking a task which he felt would agitate him . . . lest some unpleasant dispute might occur, and his conviction that if it did it would certainly prove fatal to him'.[34]

According to traditional physiology, it made perfect sense to perceive disruptions in one's emotional state as hazardous to health: the heart was, after all, at the centre of a series of humoral economies in which it was an active, rather than a passive, agent. It nurtured the 'feelings' of love or anger; it encouraged the body to respond in particular (even pathological) ways. It could be damaged by emotions, because it could be overheated by the 'hot blood' of anger, or frozen by the 'chilled blood' of grief. The heart was no less endangered according to eighteenth-century physiology, although the damage might result from hydraulic failure rather than an excessive temperature. Attempts to understand the heart as a physical organ, as opposed to a spiritual and active entity, however (leaving aside for the moment the debates by Vitalists and Animists), meant that there was greater emphasis than ever before on detecting or 'proving' that damage in material and concrete ways. Pathological anatomy, with its emphasis on cutting up and examining the body, meant that, even if emotions were transient and fleeting occurrences, their effects might be longer lasting. In place of the spiritual or immaterial causation that had provided explanatory power in earlier centuries, then, material causation became more important.

Explanations for the well-established relationship between the structure of the heart and emotional change had traditionally been found in spiritual or immaterial causes—the influence of the soul, for instance, of divine intervention, even of the balance of the constitution itself (via the humours, which were themselves mediated by the soul acting in and through the human mind).[35] In essence, this historical and mystical element granted to the actions of the heart was an acknowledgement of its spiritual status: the heart was no mere lump of muscle or organ in humoral or pre-modern physiology, but an active structure that drew and repelled blood and spirits, that guided the passions, and that mediated between mind and body as no other organ could do.

What was signified by the identification of angina pectoris as a specific and structural disease, then—a disease that operated according to a set number of predisposing causes, that possessed key characteristic symptoms regardless of individual difference, and that operated according to a specific (and usually a swiftly terminating) trajectory—was the beginning of what we might term a heart of science: a heart that was secularized and divested of any mystical or spiritual components, a heart that could be measured and weighed, its actions quantified and described, a heart that might be acted upon by emotions in ways that were barely explicable to human kind, but that in and of itself possessed no emotional qualities or potential.

Many of the assumptions that pathological anatomy set out to demonstrate (the link between emotions and heart disease, for instance), were merely giving new status to well-established beliefs. And yet pathological anatomy aspired towards newness and objectivity, to provide unquestioned accounts and comparisons of results found at autopsy with symptoms described by living patients. It was a gendered exercise that enabled the body to be viewed as a machine and necessarily secularized its workings, and impacted on emotion physiology. And emotions continued to be problematic—with their links to life, to the soul, and to the immaterial. That they were now regarded in mechanistic and chemical rather than immaterial and spiritual terms did not solve the problem of their origination, their stimulation, and their relationship with the soul.

Anatomizing the Heart

The heart that was described in eighteenth-century treatises was, as a result of intensified interest in anatomical dissection and the testing-out of observable characteristics, far more detailed and scientifically motivated than earlier. Consider, for instance, *A Compendium of Anatomy* (1739) that was published together with Alexander Monro's *The Anatomy of the Humane Bones* (1732) and included extensive reference to Monro's work:

The heart is a strong muscle, somewhat resembling a cone, or the top of a sugar-loaf, flattened on the sides, round at the top, and oval at the base. It is hollow within, and divided by a partition into two cavities, called VENTRICLES; the right is thin and soft, the left thick and solid. Each ventricle has two openings at the base . . . the RIGHT ventricle opens into the right auricle, and into the trunk of the pulmonary artery, for the reception of blood from the right auricle, and throwing it into the pulmonary artery, the LEFT into the left auricle, and into the trunk of the aorta, or great artery, for the reception of the blood from the left auricle, and throwing it into the aorta.[36]

Lacking any spiritual or immaterial function, the heart 'with its parts' was given a material and observable function as 'the chief instruments of the circulation of the blood'. The behaviour of these organs was regarded in mechanistic, hydraulic terms:

The two ventricles, like two syringes closely joined together, unite into one body, and are furnished with suckers placed in contrary directions to each other; so that by drawing one of them a fluid is let in, and forced out again by the other. This muscle is capable of two motions, viz that of contraction, and that of dilatation, which anatomists call the systole and diastole of the heart.[37]

Most accounts of human physiology paralleled this account of the heart's function, supported by pathological anatomical accounts of its dysfunction. The rise and practice of pathological anatomy as related to the heart is discussed more fully in the following chapter. In Britain it was influenced by such continental developments as Giovanni Battista Morgagni's *De Sedibus et Causis Morborum* (1761, available in English from 1769), and by the work of Xavier Bichat, Gaspard-Laurent

Bayle, and Jean-Nicholas Corvisart. Influential practitioners in England included William and John Hunter, and John's nephew Mathew Baillie (whose *The Morbid Anatomy of Some of the Most Important Parts of the Human Body* was published in 1793, and widely regarded as the first such textbook on human anatomy).

Of these works, that of Morgagni, in particular, provided the earliest foundations for a structural and organic understanding of diseases like angina pectoris. Morgagni's *De Sedibus* made use of over 700 case reports in 70 letters with reference to many other authors, and nearly 500 of Morgagni's case studies could be said to have shown symptoms of angina pectoris.[38] The comprehensiveness and consistency of Morgagni's approach, moreover, provided a pointer to the systematic clinico-pathological correlation that would become so widespread in the nineteenth century, and in the work of Jean Nicholas Corvisart and others.[39]

Dissecting Angina Pectoris

Identifying the nature and structure of changes in the heart was one thing; proving and understanding causation was something else. Despite the widespread acceptance of the view that emotions caused heart disease in general, and that the passions were one of the main causes of angina pectoris, there was some disagreement over the precise changes that accompanied the disease. These were summarized by James Hope in his influential *Treatise on the Diseases of the Heart and the Great Vessels* (1832):

Great diversity of opinion has existed respecting the cause of angina pectoris. Different physicians have found it connected with different organic lesions or states, and each has supposed it to be occasioned by that, with which he has most frequently found it coexist. Dr Parry, and after him Burns and Kreysig, ascribe it to ossification of the coronary arteries; Dr Hooper to affections of the pericardium; Dr Hosack to plethora; Dr Darwin to asthmatic cramp of the diaphragm; Drs Butler, Macqueen and many others, have regarded it as a particular species of cramp; Dr Latham has found it connected with enlargements of the abdominal viscera while the thoracic viscera were sound; and Heberden, having found it both connected and unconnected with organic disease, thinks that its cause has not been traced out, but that it does not seem to originate *necessarily* in any structural arrangement of the organ affected.[40]

By the time of Hope's writing, it was not considered necessary to conclude that any one disease was to blame. The newly emergent concept of 'nervous' disorders of the heart meant that the heart could be functionally and structurally prone to disease if the heart was 'irritated' or 'morbidly susceptible of irritation', a condition commonly associated with structural disease.[41] By the nineteenth century, then, far more emphasis was placed on functional changes—in part, the belief that emotions, by disrupting the circulation and the blood flow, could cause changes in the heart's action, even if the heart itself was unchanged. Not until the twentieth century would functionalism be recast as neuroses.[42]

We have seen how Hunter's condition was decided upon at autopsy, the structural ossification that was understood to be a common cause of angina pectoris. We have also seen how the acquisition of medical knowledge during the eighteenth century came increasingly to rest upon the ability to compare conditions found at autopsy with lived and observed experiences on the bodies of patients. By the nineteenth century a whole other level of awareness—that of new technologies designed to probe into the cavities of the living subject—would be called into question. For now, however, what was important was the ability to make parallels between the objective apprehension of a patient's symptoms and the physical evidence of those symptoms when revealed to the anatomists' gaze.

The Anatomist as Case Study

It is understandable, therefore, as part of this peculiar construction of medical knowledge and legitimate experience, that the case study came to occupy the central status that it did in the accounts given by eighteenth-century writers on heart disease and angina pectoris. It is also understandable that the body of John Hunter itself became an important symbolic and practical touchstone for the course of angina pectoris as a disease. From his passionate temperament, through his lack of self-control, to his experiences of cardiac illness and subsequent death, the anatomist became a case study in his own right—an exemplary model of angina pectoris writ large for the medical profession to observe and to learn from.

The first observed account of Hunter's illness took place in the spring of 1775, a full eighteen years before his fatal attack at St George's Hospital. The evidence of something untoward in Hunter's health was 'an alarming attack of spasm' that was attended with a 'cessation of the heart's action'. This is understood to have lasted almost 'an hour in duration' and 'in defiance of several active remedies suggested by Dr [William] Hunter' and at least three other physicians who were immediately summoned.[43] This apparently marked the first 'warning' that Hunter received about possible disease of the heart, though it was not until many years later that the 'organ became permanently deranged'.[44] Hunter's acquaintances, friends, and colleagues monitored his condition with interest. In a letter of 1785, written when Hunter had just left for the spa waters of Bath to recuperate, the physician and philanthropist John Coakley Lettsom commented that he had recently called Hunter to examine one of his patients. Hunter was 'going from this busy stage', Lettsom remarked (though whether of life itself or of Hunter's involvement in the London medical scene is uncertain), 'he can scarcely go up stairs so much is he affected with dyspnoea on the least motion'.[45] Another friend of Hunter's, the physician Edward Jenner, similarly expressed concern about Hunter's heart, writing to William Heberden that he was 'fearful if Mr H. should admit this, that it may deprive him of the hopes of a recovery'.[46] Even the fear caused by possessing angina pectoris could prove fatal.

By 1785, these attacks of chest pain and disruptions of the heart's action had become commonplace for Hunter, 'especially after an occasion of any extra exertion or mental anxiety'.[47] The psychological origin of Hunter's physical deterioration was often made explicit. Whether exacerbated by gout or environmental factors, it was his temper, or 'a violent mental affection', that proved to be 'the immediately exciting cause'.[48] This was the case each and every time Hunter became subject to an attack of spasm about the chest, as well as in the originating weakness of the heart itself. Two years after his initial 'warning', for example, Hunter experienced an 'attack of illness' caused by 'anxiety of mind at being called on to pay a large sum of money for a friend, for whom he had become security'.[49]

Hunter's heart became part of his mythology even before his death, as Hunter's condition was talked about and considered by his friends and colleagues. Those views even made it into works on angina pectoris. In Parry's *Inquiry*, for instance, the thoughts of Edward Jenner on Hunter's condition were considered at length in the introduction, largely in order to demonstrate the structural changes wrought by the disease. In a written response to Parry, Jenner wrote that he had begun to suspect angina pectoris was a result of coronary artery disease after his second dissection:

I was making a transverse section of the heart pretty near its base, when my knife struck against something so hard and gritty, as to notch it. I well remember looking up to the ceiling, which was old and crumbling, conceiving that some plaister had fallen down. But on a further scrutiny the real cause appeared: the coronaries were become bony canals. Then I began a little to suspect. Soon afterwards Mr PAYTHERUS met with a case. Previously to our examination of the body, I offered him a wager that we should find the coronary arteries ossified.[50]

Jenner lost his bet, although the arteries were found to be hardened. It was about this time, he recalled, that his 'valued friend, Mr John Hunter, began to have the symptoms of Angina Pectoris too strongly marked upon him; and this circumstance prevented any publication of my ideas on the subject, as it must have brought on an unpleasant conference between Mr HUNTER and me'.[51] He did mention his thoughts to Hunter's supporter and fellow surgeon Henry Cline, however, and to Everard Home, Hunter's brother-in-law and fellow surgeon. When Hunter died, and Home performed the autopsy, he wrote to Jenner to tell him that he had been right.[52]

The extent to which physicians were detached observers of their own conditions (or even used those conditions to further their knowledge of a field) is a subject that is increasingly considered by medical historians.[53] In the case of Hunter, his controversial and arguable self-infection with venereal disease has occupied rather more historical attention than his cardiac symptoms or his emotional state.[54] Part of his professional identity construction—as rational scientist, on the one hand, as temperamental genius subject to such extreme emotional states, on the other—was summed up by Palmer in his 1835 edition of the *Works of John Hunter*: 'a painful thought that one possessing a mind of such intellectual vigour should, from neglecting earlier to check this infirmity of temper, at length have allowed it "so to over-master reason" as to reduce him to hold his life on such a tenure'.[55]

Hunter's own autopsy of Heberden's patient initially suggested that he did not believe that there was any physical or organic basis for

angina pectoris, a claim subsequently challenged by Edward Jenner.
Again in 1776 Fothergill gave a 'Farther Account of the Angina
Pectoris', in which he discussed the case of Mr Rook, a 54-year-old
man who died unexpectedly 'in a sudden and violent transport of
anger'. In this instance, as in 1772, John Hunter was asked to 'open'
the body, and he reported that it did show evidence of the ossification
of the coronary arteries.[56] There is no evidence, however, that Hunter
was conscious of, or believed in, the same process occurring within
his own breast. Although Hunter did not perceive any specific organic
changes being related to angina pectoris, at least that are detailed in
the surviving papers, he was aware of his own position as a subject.
He was also aware of the importance of his own position as a sufferer,
having asked that his heart be retained for examination after his death.
Unfortunately, Everard Home did not honour this request.

Emotions and Vitalism

Part of the reason why establishing causes of angina pectoris was so
problematic to eighteenth-century medics as a philosophical as well
as a physiological issue was that it drew attention to the ability of
emotions, as elusive, immaterial phenomena, to disrupt the fabric
of the physical body and the carefully laid-down constructions of
that body that had emerged since the later seventeenth century. The
emotions occupied an ambiguous position as potential evidence of
the existence of a soul, a vital force, something that could not be
accounted for in materialistic terms.

Central to the link between agency and the heart was the theme
of emotions, the passions, elements, or components that apparently
drove the individual to act, that acted as motivation. Understanding
the impact of emotions on the heart, therefore, seems to strike at the
heart of some of the most important theoretical issues in eighteenth-
century physiology: that is, the extent to which physical and material
structures acted independently and without the need for intervention
from any innate principle, such as a soul.

This would become a central theme in the work of the Cambridge
physiologists in late Victorian Britain, as the problem of the heartbeat
dominated the work of Michael Foster and his research school.[57]
As in previous centuries, the heart continued to be at the centre

of debates on the mind–body relationship, the emotions, and the soul. The immediate context in which Hunter was working was, therefore, one with considerable challenges for the anatomist. Where was the soul? What were the limits of its influence? If the body had a soul, many mechanists maintained, it must be a rational, immaterial principle that was somehow attached to the material body, and operating in accordance with established laws on matter and emotion.[58] Many vitalists and animists rejected this limited status granted to the soul, claiming that, rather than being an adjunct of the body (its physiological role restricted to 'willed activity and consciously perceived sensations'), the soul or *anima* acted independently.[59] Indeed, the soul performed the ordinary functions of life in humans, although those of lower animals might be performed by mechanical principles. For the animist Georg Ernst Stahl (1659–1734), court physician to Frederick William I and professor at the University of Halle, 'the soul' (a term he used to cover *all* perceptual processes, sensory, mental, and emotional) formed images and ideas, incited affective responses, and effected physiological change within the body. Stahl's study of the physiological impact of emotional responses led him to assert that the body followed the soul, an idea that could not be *disproven* medically, as the actual relationship between emotional experience and expression remained uncertain. Moreover, in rejecting the idea of the body as an automaton, Stahl regarded it as an 'inherently unstable instrument of life, constantly in need of regulation and repair by the soul'.[60]

How far the emotions acted as a result of the soul's influence, or—more prosaically—as a result of mechanistic forces and the ebb and flow of the human fluids/body, was implicit if not explicit in the work of many physiologists and morbid anatomists, Hunter included. In his work on physiology, for instance, Hunter stressed his belief in some principle that was not reducible to structural or material components, a 'living principle' that was not 'in the least mechanical, nor is it in the least connected with any mechanical principle'.[61] In the following century, this question would be resolved, at least in part, with the removal of emotions to the brain, and away from the heart. The question of the heart's irritability or otherwise would become central to experimental physiologists in the nineteenth century, albeit largely divested of its prior psychological and emotional importance.

That, at least, is the theory. But what had really changed in eighteenth-century experimental work into the heart and its role as an organ of emotion? In some way, showing the status of the heart of angina pectoris, a disease that took place in the heart largely as a result of emotional factors, shows the retention of those links between mind and body, between emotions and the heart, that were increasingly downplayed by scientific materialism. And yet the new stories and categories of explanation that were emerging arguably did little more than redraw new theories along old lines, by asserting the connectedness of mind and body and of the heart as the centre of the lived experience of emotions. There was, therefore, an ambiguity about newly emergent categories of heart disease, which struggled to reject the holistic framework in which the heart was traditionally understood.

By way of example, let us return to the body and the heart of John Hunter. We have seen that, whatever importance was placed on the heart as a material organ by him and by other anatomical specialists in Britain and continental Europe, the heart remained an organ of emotion as well as the body. It was impacted upon by the moves and moods of an individual's experience, and for good or ill. In this the mind and body were linked, a position made possible by nervous physiology that united the experiences of each through the doctrines and notions of sensibility and sympathy.

Hunter's own research into the nervous system stressed this interconnectedness, and the two-way flow between mind and body:

The actions of the brain towards the body are...of two kinds: one in consequence of the feelings or state of mind at the time (actions of fear, courage, anger, love, etc.)...the other...the command of the will, called voluntary actions.[62]

Each of these actions takes place through the nerves, and the influence of the mind over the body (and vice versa) was indisputable. There is not, he declared, 'a natural action in the body, whether involuntary or voluntary, that may not be influenced by the peculiar state of mind at the time'.[63] Hunter's belief in the psychosomatic nature of illness was carried into his practice—giving bottles of 'common hot water' to his wife under pretext that it came from Bath spa, and she deriving 'all the benefits from them that she had expected from a course of the real waters'—and in his research into hysteria, somatoform disorders, and hypochondriasis.[64]

The Ambiguity of Angina Pectoris

The ambiguity of angina pectoris, as a product of 'old' and 'new' discourses on the body and the heart, would become more apparent in the nineteenth century. On the one hand, angina pectoris represented the continuation and preservation of traditional ideas about the heart and its influence by a variety of emotional, psychological, and environmental factors. On the other hand, it was a 'new' disease, a specific set of symptoms that could be measured and compared and that presented a clear path from diagnosis to prognosis (and usually) to death. This ambiguity was apparent both in the treatment of the heart as an emotional organ (and subject through palpitations and irregular movements of the pulse, for instance, to the emotional experiences of individuals), and also as an organ of science: a material body subject to decay. It was also apparent in the recommendations and therapeutics made by physicians towards the cure of angina pectoris.

In traditional understandings of the body, continued into the eighteenth century by physicians like John Fothergill, 'vehement emotions' were dangerous to the health in general and to the heart in particular. For 'excesses of passion and anxiety . . . contribute more to the increase of [heart disease] than a combination of all the other causes'.[65] Like the rest of the 'non-naturals'—those regimental habits that included sleep, exercise, diet, air, excretions, and passions of the mind—balance and moderation in emotions was necessary for a peaceful and healthy existence.[66] Although, at the time of Fothergill's writing, fluid-based conceptions of mind-body interaction were outmoded in favour of hydro-dynamic or nervous theories, which emphasized the solids, rather than the fluids, of the body, the symbolic and physical status of the heart as mediator of the passions remained intact.[67] Preventing and treating heart disease in the eighteenth century continued to focus primarily on maintaining an emotional and physical equilibrium.

For all his acclaimed status as a man of genius, a man of science, then (or perhaps because of it), Hunter personified the wrong living of the time. He signified the archetypal angina-pectoris patient because he was middle aged, because he was male, because he experienced tempestuous emotions, and because he was an excessively hard worker who did not eat properly or get sufficient sleep.[68] Discussion of all these extremes—and of their accommodation into a discourse on hydraulics, blood flow, blood pressure, and so on (in place of humours and the soul)—helped to construct a scientifically viable explanation of the relationship between emotions and the heart (and between emotions and the body) that rewrote humoral principles along new lines.

Unsurprisingly, therefore, medical therapeutics changed little as a result of new and materialistic disease categories. All medical writers, regardless of whether they believed angina pectoris to be structural or functional, highlighted the importance of 'right living'. Parry elaborated how the patient must not get too much heat—'whether of the sun or fire'—and must avoid stimulating cordials in diet or medicine. Exercise was important for keeping up the circulation, for keeping the blood moving, lest inaction, 'stooping', or 'long sitting' pressed too heavily onto the femoral arteries, preventing 'the flow of blood to the extremities' and allowing it to 'accumulate in the heart and larger vessels'.[69] Of course, it was equally important not to do too *much* exercise, or to allow the heart to be moved by anger and its 'fatal effects'. Here, as in most areas, Parry advised, it was necessary for the patient to 'minister to himself; and in this country, in which we have reason to thank Providence that there is still some genuine religion, it may be hoped that the sufferer will not be at a loss for sources from which to derive forbearance and consolation'.[70] At the end of the eighteenth century, as at the beginning, religion was still the best recommendation that physicians could make for reining in the passions, and for making them moderate.

3

Knowing the Heart: From Morbid Anatomy to New Technologies

By tracing the analogies that exist between diseases as they occur in different parts of the body, not only in the study of Pathological Anatomy much facilitated and abbreviated . . . we are greatly assisted in acquiring an accurate knowledge of diseases, of perceiving differences in their general character, and of establishing general principles for their treatment.

Matthew Baillie, Morbid Anatomy (1793)[1]

At the moment of the arterial pulse, the ear is slightly elevated by an isochronous motion of the heart, which is accompanied by a somewhat dull, though distinct sound. This is the contraction of the ventricles. Immediately after, and without any interval, a noise resembling that of a valve, or a whip, or the lapping of a dog, announces the contraction of the auricles. (I make use of these trivial expressions because they appear to me to convey better than any description, an idea of the nature of the sound in question.) This noise is accompanied by no motion perceptible by the ear, and is separated by no interval of repose from the duller sound and motion indicative of the contraction of the ventricles, which it seems, as it were, to interrupt abruptly.

René T. H. Laennec, A Treatise on the Diseases of the Chest (1827)[2]

Between the eighteenth and nineteenth centuries, diagnoses and classifications of heart disease underwent a profound change in Britain, both in terms of disease concepts, and in the role of the heart as an organ of disease. If we look at the two quotations used in the epigraph, we can see an immediate difference in diagnostic categories. In the first example, Baillie's *Morbid Anatomy*, the heart can be analysed and understood, its changes seen, only after the death of the patient. In the previous chapter we saw how case studies, such as that of John Hunter, were constructed through a series of pathological anatomies. By enumerating and comparing disease structures in the bodies of dead patients, it became possible for physicians to 'perceive differences' in the character of diseases and to establish 'general principles' for therapeutics (see Fig. 4).[3]

By the early nineteenth century, as seen in the work of Laennec, greater emphasis was being placed on understanding the hearts of the living through a series of innovative diagnostic techniques. In the case of Laennec, as will be seen, that was through mediate auscultation—the precise measurement and assessment of heart sounds detected through the stethoscope. A series of elaborate tropes—the heartbeat being compared to a 'valve', a 'whip', or the 'lapping of a dog'—was constructed to distinguish between types of detectable diseases.[4] Mediate auscultation was the first in a series of diagnostic tools to be developed over the course of the nineteenth century.

This chapter analyses how these new sensory interventions or technologies helped to shape theories of heart disease through a literal and figurative lens of objectivity and distance. Tracing the development of early diagnostic strategies—including auscultation and percussion—it shows how medical specialism before the mid-nineteenth century focused increasingly on the functions of the heart as measurable, quantifiable, and comparable. Those processes of measurement and quantification formed part of a broader process of objectification that distanced the physician from the patient and reduced the role of subjective analysis in the measurement of cardiac function.

Unlike those of the later nineteenth century, however, early innovations continued to emphasize the heart's role as an integral part of the body system, impacted on by the emotional, constitutional, environmental, and regimental world of the patient. Emotion remained central to the heart of disease in physical, tangible ways: extremes of anger or anxiety could produce structural changes in the heart, including lesions and arterial obstruction. Evidence of such change came not only from clinical diagnoses of the living heart, but also

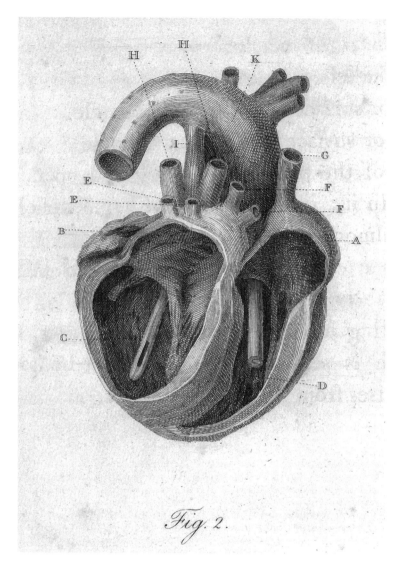

Fig. 2.

Fig. 4. Matthew Baillie and the heart of disease.
The heart anatomized by Matthew Baillie. One of a series of
engravings published in 1799.
Source: Wellcome Images.

through the comparison of the sounds and movement of that heart
with clinico-pathological correlation conducted at autopsy. Before
analysing the development of new technologies by which heart functions were mediated, therefore, we need to consider the development
of heart-disease concepts in the context of morbid and pathological
anatomy.

Morbid Anatomy and the Heart of Disease

The development of morbid, or pathological, anatomy in the scientific
medical tradition is a lengthy and diverse subject; some evidence of it
working in action has already been considered in the previous chapter
in the case study of the surgeon (and pathological anatomist) John
Hunter. As Russell C. Maulitz has recognized, a full account of its
development would need to include pathological museums and the
collecting impulse, as well as the development of disease concepts and
their localization in individual bodies.[5] That is obviously beyond the
scope of this chapter. What was important for the early construction
of disease concepts in the history of the heart, however, was the
emergence of anatomical dissection for teaching practices, and the
subsequent influence of those findings on diagnostic strategies and
therapeutics.

As detailed in the previous chapter, the Italian anatomist Giovanni
Battista Morgagni made one of the first systematic attempts to codify
organ pathology.[6] This marked the emergence of anatomically localized interpretations of disease, rather than the humoral or holistic
enterprise that had previously been the physician's lot.[7] This enumerating of disease concepts within individual organs of individual
bodies took place a generation before the development of pathological anatomy in the influential French medical tradition.[8] It also
provided a pointer to the systematic clinico-pathological correlation
that would become so widespread in the nineteenth century, and
in the work of the French physician Jean Nicholas Corvisart, discussed in more detail below. Above all, pathological anatomy aspired
towards objectivity, the principles of which, along with the way
that objectivity shaped scientific identities and communities, have
recently been explored by Lorraine Daston and Peter Galison.[9] Part
of that shaping was the distancing—psychological and physical—of

the physician from the patient, a crucial stage in its construction as scientific knowledge.

Influential practitioners in England who were influenced by Morgagni included William and John Hunter, and John's nephew, Matthew Baillie, each of whom we met in the previous chapter. Hunter's own heart dysfunction contributed to the development of theories of heart disease, as discussed in the previous chapter. Baillie's *The Morbid Anatomy of Some of the Most Important Parts of the Human Body* (1793) was produced in eight English, three American, and five foreign-language editions.[10] Before Rokitansky's *Pathological Anatomy* (1842), it was arguably the most consulted textbook on disease changes in the human body.[11]

Making use of John Hunter's collection of wet specimens, as well as of dissections that he had personally undertaken, Baillie built on the earlier work by Morgagni and listed the numbers and types of disease formations that might be identified in the organ of the heart. These included 'older' diseases such as 'polypus'—rejected by the 'moderns' as a disease, and characterized by an accumulation of 'lymph' in the heart's chambers.[12] More recent formulations included a range of aneurisms and valvular blockages, the interpretation of which took for granted the mechanistic and hydraulic function of the heart and the circulatory system. The hearts described in eighteenth-century treatises like Baillie's were necessarily far more anatomically detailed and precise than those in seventeenth-century texts. Their function was explained through mechanistic, hydraulic language:

The two ventricles, like two syringes closely joined together, unite into one body, and are furnished with suckers placed in contrary directions to each other; so that by drawing one of them a fluid is let in, and forced out again by the other. This muscle is capable of two motions, viz that of contraction, and that of dilatation, which anatomists call the systole and diastole of the heart.[13]

Most early nineteenth-century accounts of human physiology reproduced this account of the heart's function, supported by pathological anatomical accounts of its dysfunction: the heart was a material structure or pump. The influence of the heart was initially direct and mechanistic (by the end of the century it would be primarily nervous or sympathetic), and a range of structural and immaterial influences impacted upon its action. From the late eighteenth century there were several attempts to categorize individual heart diseases according to localized lesions. After Morgagni, the only significant textbook on cardiovascular disease was Jean-Baptiste de Senac's *Traité de la structure du cœur*, a work published in 1749.[14] Senac's work provided a detailed account of the structure and function and anatomy of the heart, and noted the increasing incidence of cardiac problems with age. He also discussed the difficulties of diagnosing diseases of the heart, the key features of which ranged from palpitation and asthma to oedema (characterized by fluid accumulation in the abdomen) and haemoptysis (coughing up blood).[15]

After Senac, other medical theorists described rheumatic disorders of the heart and congenital forms of heart diseases. These findings were gradually encouraged, and spurred on, by a range of diagnostic techniques, from percussion and auscultation to the use of the stethoscope and later more technologically driven devices. Gradually, however, the emergence of the heart of disease gave rise to a far more elaborate system of identification and symptomology than had previously existed.[16] By the early nineteenth century, several forms of heart disease were identified as medical entities. The anatomizing of the qualities of the heart in health and disease facilitated this trend, along with a more general drive towards standardization of medical techniques.[17]

The earliest development of techniques to 'know' the heart included the adaptation and application of traditional, subjective skills on the part of the physician, working in conjunction with the patient, to listen to and understand the quality (rather than the quantity) of the heartbeat. The most commonplace of these were the combined practices of auscultation and percussion. Later innovations would include the sphygmograph, the sphygmomanometer, the electrocardiograph, and the X-ray, each of which has been discussed in technological terms in histories of cardiology.[18] In many cases, there was a lag between the emergence of experimental technologies in the laboratory and their application in diagnostic practice.[19]

Tapping and Listening

Auscultation, the practice by which the physician listened to the chest of the patient with the unaided ear, was commonplace in Europe for centuries. It was generally accompanied with percussion, the physician tapping on the chest of the patient to detect abnormalities in the sound produced by the cavity. This methodology of the procedure probably increased in rigour following Jean-Nicolas Corvisart's translation of Joseph Leopold Auenbrugger's *Inventum Novum* in 1808—a work that was popularized in Britain by John Forbes's translation of René T. H. Laennec. Laennec's own extensive influence on cardiac diagnostic practice is discussed in more detail below.[20]

Joseph Leopold Auenbrugger (1722–1809) was an Austrian physician who identified the different sounds that came from across the wall of the chest when it was tapped by the physician. In 1761 he published these findings in *Inventum Novum ex Percussione Thoracis Humani ut Signio Abstrusos Interni Pectoris Morbos Detegendi* (*New Invention, by Means of Percussing the Human Chest, as a Sign of Detecting Obscure Diseases in the Interior of the Chest*).[21] By systematic percussion, Auenbrugger identified the boundaries of the lungs, and detected when the resonance of any area was diminished because of fluid build-up. The popularization of Auenbrugger's technique was effected largely through the influence of Corvisart (1755–1821), physician at the Parisian Hôpital de la Charité from 1785, and later personal physician to Napoleon Bonaparte. Corvisart was a transitional figure in the shift from anatomic–pathologic medicine to physical diagnosis. As a teacher, he influenced the work of future physicians like Laennec, clinicians such as Bayle and Bretonneau, and anatomists like Cuvier.[22] He also helped shape and develop a number of cardiac disorders.[23] In *An Essay on the Organic Diseases and Lesions of the Heart and Great Vessels* (1812), Corvisart recommended the use of percussion in the context of detailed case-study evidence of individual patients, which were then compared to anatomical findings at autopsy.[24]

This correlation of symptoms and structures between the living and the dead was part of a more general shift away from understanding disease as a constellation of subjective symptoms, and towards a nosology of disease classifications and types. It is also illustrative of a move towards an understanding of disease as linked to particular organs of the body, rather than the body system as a whole. However, that did not mean—as it would in later decades—that the organs should be considered outside the context of bodily and mental experience. For a holistic understanding of the body and its mutual influences with the mind meant that reshaping of the human heart in health and disease initially conformed to much earlier assertions about links between emotions and the heart. As is apparent in Corvisart's writings on the heart, the organ remained embedded in a series of social and psychological (and even spiritual) networks that were not reducible to material structures.

'Detectable Signs'

Corvisart's *Essay* reminded readers that the application of physiological anatomy to understandings of disease processes was not best served by 'observations made on bodies disinterred, and taken at random into the public amphitheatres'. Rather, what was required was 'to study on the living and diseased man, the characters peculiar to lesions of the different organs, to observe well their phenomena, and establish their symptoms, by observations sufficiently numerous, to prevent the possibility of misunderstanding them'.[25] Corvisart took up this theme in his introduction to the volume, as:

The more exact anatomy is studied by physicians, the sooner they will be able by careful observation to distinguish and establish among diseases, a great number of organic lesions, whose existence has not by a majority of them been even suspected. [Emphasis in original][26]

For Corvisart, satisfactory attention to the study of the 'living man', to the 'numerous sensible phenomena of the actions of the parts', meant that the physician

must sedulously apply himself both to the physical and moral man and (except the bond which unites this double being and which has ever been concealed from the human eye), the physician must perceive the most delicate perceptible influence of the reciprocal action of the one over the other.[27]

The 'passions and vices' of the individual were, therefore, of as much relevance to the physician as the movement of his or her muscles and limbs, for these 'animate, excite, stimulate and convulse the whole system, in a thousand different ways'.[28] The 'moral affections', of which emotions were included, were seen to have 'a powerful influence on the evolution of diseases, and particularly on that of the organic lesions of the heart'.[29]

This organ is the point in which the effects of all the moral affections, gay or melancholy, seem to be concentrated. No moral affection can be experienced, without the acceleration, diminution, or derangement of the motion of the heart; let its power be increased, paralysed or destroyed, pleasure, pain, fear, anger, all the sensible affections, in fact, make it palpitate, or suspend its action.[30]

There were, according to Corvisart, many 'signs' the physician could use in order to diagnose organic heart disease in the living. These included:

1. The expression of countenance, the external state of the body and the means which may be externally employed for the purpose of becoming acquainted with the diseases of the heart.
2. The various derangements observed in the circulation.
3. Such as happen in respiration.
4. The state of digestion.
5. The influence of the affections of the heart on the secretions and functions of the cerebral Organ [the brain].[31]

First on the list of 'detectable signs' the physician should look out for was the 'figure, physiognomy and the *facies propria*', or external appearance of the patient. In 'persons of a sanguine temperament', Corvisart recommended, his language reflecting humoral influence, there was a 'sudden and transitory redness of the face'. In women especially, this was attended 'with laborious respiration, and slight and frequent palpitations'.[32] When the disease progressed, the *facies propia* became more 'expressive':

In general, the countenance becomes bloated, it is *vultuous*, but not exactly as in acute diseases; the size of the face is even greatly enlarged, but less discomposure is observed, and less alteration in the features. The countenance is generally of purple color; the whole venous system may be said to be injected. The lips and nose present this purple or violet tinge in a more striking manner.[33]

The examination of the rest of the body was as important as that of the face, such as the engorgement of the jugular veins, their pulsation, the irregularity of the pulsations, 'a sort of rushing' felt through the body, and the swelling of the extremities.[34]

Next on Corvisart's list were circulatory 'derangements', often 'exhibited externally by phenomena sensible to the sight or touch'.[35] These could be detected by administering to the measurement of the pulse at various points, but it was important to distinguish between these apparently objectively perceived measurements and the subjective 'palpitations of the heart, its contraction and trembling, which are sensible only to him who experiences them'.[36] For the most objective measurement possible, it was, therefore, necessary to examine between palpitations, or to attempt to make allowances for them.

Respiratory problems occurred when the heart complaint became advanced. There were commonly 'slight, but habitual' difficulties in breathing at first, and respiration became 'embarrassed, high, short and interrupted'.[37] In addition, amongst aneurisms of the aorta there is often a 'sort of hissing' that develops in the final stages. There is no position in which a patient afflicted by respiratory difficulties may obtain relief, so fully is the 'natural organization' of the heart and the vascular system affected.[38]

In a final section dealing with 'digestion, secretions, and functions of the brain', Corvisart referred back to his section on aneurisms. Digestion became disordered, with some patients being 'incessantly tormented with hunger', and constipation was common. 'Secretions and exhalations' included the passing of 'brick-coloured' urine and an eventual 'effusion' in the abdominal cavity.[39] The disturbances to brain function ranged from dizziness and syncope to delirium and despair.[40] It was unclear how far 'moral affections' could produce such varied and physical symptoms. In the early stages of aneurism, causes could include 'a lively and protracted moral affection' or 'a menstrual or haemorrhoidal suppression', which before long led to 'palpitations' or 'pains in the region of the heart':

Can it then be pronounced that the person, threatened with these symptoms, is affected with an aneurism in the heart in the first period? Certainly, such a decision would be too inconsiderate. Hence, let the moral affection cease and the suppressed evacuation be immediately re-established, and every symptom of the disease will at once disappear.[41]

Like 'suppressed evacuation', then, 'moral affections'—extremes of emotion—could produce functional and structural problems in the heart that, unless alleviated swiftly, could become fatal.

Corvisart's work in the history of cardiac diagnosis has been over-shadowed by that of Laennec, whose *Treatise* was published two years before Corvisart's death.[42] Yet much of Laennec's writing borrowed, with acknowledgement, from the work of Corvisart, and his findings helped shape the 'science and art' of clinical diagnosis of the heart and chest, primarily by popularizing the art of *mediate* auscultation (the mediating principle being the stethoscope).[43] By early 1817 the stethoscope was being used for cardiac auscultation, for the detection of problems within the heart itself.[44]

Laennec's *Treatise* was initially published as a contribution to pathology, though its physiological relevance was recognized.[45] Perhaps unsurprisingly, given the relationship of the two authors, the link between the emotions and the heart was as strong in Laennec's work as in Corvisart's. Duffin has estimated that, of the fifty complete case histories dealt with in Laennec's *Treatise*, at least fifteen displayed psychological preconditions along with physical ailments. Examples include those who had been subject to psychic distress or who had experienced extreme fright, or grief at the death of a spouse.[46]

Where used in clinical practice, the stethoscope was most useful in detecting changes in the heart's shape and bulk. The recognition of such changes was consistent with, and influenced, the language, structure, and onset of heart-disease symptoms. Before the nineteenth century, the phenomena most frequently associated with heart disease were palpitation, chest pains, and sudden death.[47] The diagnostic use of the stethoscope expanded (and identified) the categories of symptoms associated with cardiac disease, especially when it supplemented more traditional analyses of the complexion, the constitution, and the pulse.

Rationalizing the Pulse

From the 1820s, physiological experimentation in France focused attention on the pulse and arterial motion, with the heartbeat posing as many questions for experimentalists as it did for philosophers. Along with auscultation and percussion, the measurement of the heartbeat via the pulse was the most common way of understanding the functions (and dysfunctions) of the living heart before the mid-nineteenth century. That did not mean, however, that interpretations of the pulse (or pulse*s*) were constant. In terms of heart function and diagnostic practices, the pulse had been linked to the health of the heart in a variety of ways, as well as to the emotional experiences of individuals, since the time of Galen.[48]

By the late eighteenth century, then, there was already an established tradition of measuring and interpreting the pulse as a sign of cardiac activity and health, as well as the emotional equilibrium of the patient. The earliest known testimony originates in the work of the Chinese physician Pien Ts'Io, who lived and worked around the fifth century BC, and is credited with recognizing the pulse as a diagnostic tool.[49] In ancient China, analysing the pulse involved examining eleven different parts of the body. Six different pulses were detectable in the wrists alone, and each pulse corresponded to a separate organ.[50] There were around 200 pulse variations that the physician had to keep in mind, the individual meanings of which determined diagnosis, prognosis, and treatment.[51]

Although measuring the *number* of beats per minute had been in common usage since at least the seventeenth century, before the mid-nineteenth century it was the qualitative, rather than the quantitative, measurement of the pulse that mattered. Interest in the pulse peaked after the popularization of theories of blood circulation, only to be revived, along with other aspects of 'pulse-lore', in the eighteenth century.[52] Yet it was not until the twentieth century, with the Burnley physician Sir James Mackenzie's *Study of the Pulse* (1902), and the development of the electrocardiograph, that feeling the pulse was standardized and professionalized as an aspect of Western medical practice.[53] Prior to this period, its qualitative characteristics, and the interpretative perspective of the patient, remained relevant.

In his 'General Remarks on the Practice of Medicine' (1863), the physician Peter Mere Latham described not only the pulse's number, frequency, succession, and regularity, but also its relational characteristics: 'hard', 'soft', 'large', and 'small'.[54] Latham allowed for some individual variation in the pulse. Yet there was a distinctive shift in medical writings over the next few decades towards establishing a standard measure. This shift was influenced largely by a move towards quantity rather than quality in reading the pulse, and by the emergence of electro-cardiographical measurements to provide efficient levels of accuracy. The application of such methods suggested that the actions of the heart were no longer measured according to the patients' subjective experiences (whose complaints of irregular heartbeat, for instance, could be tested and verified, or rejected), or even by the sensory perception of the physician. Knowledge of and control over the diagnosis of the pulse was passed instead to an apparently objective arbiter. Several new 'pulse disorders' began to emerge during the late nineteenth century, as measurements of its movements became more precise and deviations more pronounced.[55]

Mapping the internal movements of the body, and even narrativizing those movements into visual forms, was characteristic of the nineteenth century's more technological diagnostic innovations. These included the sphygmograph and the sphygmomanometer (both used to measure the arterial pulse from the mid-1880s), and by the turn of the century the electrocardiograph and X-rays.[56] The development of these precision apparatuses reflected a wider drive towards precision in the biological, physical, and chemical sciences and not just medicine. The capacity for precision, in turn, affected the kinds of measurements that were sought. Thomas Henry Huxley, the English biologist, made this point as early as 1876:

The conception of the problems to the investigation of which such apparatus is applicable was impossible until the physical and chemical sciences had reached a high degree of development, and were ready to furnish not only the principles on which the methods of the physiological experimentalist are based, but also the instruments with which such inquiries must be conducted.[57]

It is important to be wary of the technological determinism apparent in some historical accounts; all physiological innovations were subject to a degree of politicking and suspicion. There was also a significant disjuncture between innovation and implementation in medical practice.[58]

One textual corollary of the expansion of interest in diagnostic science was a proliferation of medical testimony expressed through case histories, medical guides, and textbooks. It has been stated, above, that there was a lack of writing about heart disease in the eighteenth century aside from the texts by Morgagni and Senac. Corvisart's *Treatise* proved extremely influential in rectifying this lack: within a decade of its publication, four separate textbooks on the heart were produced in Italy, Germany, America, and Britain.[59] The British and American contributions were both published in 1809; these were the Boston-based surgeon John Collins Warren's *Cases of Organic Diseases of the Heart* and the Edinburgh anatomist and pathologist Alan Burns's *Observations on some of the Most Frequent and Important Diseases of the Heart*.[60] Warren had attended Corvisart's lectures in Paris, but Burns was apparently unaware of Corvisart's work.[61] These were the first in a series of textbooks to blend clinical diagnosis and pathological investigation and depict the heart as a complex anatomical muscle. Each gave an outline of the heart's physiology, followed by sections on the most common forms of heart disease.

In 1828, for instance, the English physician James Hope returned from clinical work in Paris to establish his own practice in a fashionable area of London. He championed the use of Laennec's stethoscope, and explicitly attempted to incorporate the various strands of clinical diagnosis, clinico-pathological investigation, and pathophysiology into his study of the heart and its diseases. In 1832 he published *A Treatise on the Diseases of the Heart and Great Vessels, Comprising a New View of the Heart's Action*, a work that was widely acclaimed and went through three editions during his lifetime. This treatise was based on 1,000 case histories, and received commendation by many other cardiac specialists—including 'Heart Latham'—for its ability to record graphically, through a series of carefully detailed plates, the functions of the heart and its diseases.[62]

From 'Functional Disease' to 'Cardiac Neurosis'

In works like Hope's *Treatise*, the heart was depicted as a complex yet comprehensible organic structure. Though a structure of the body, it was influenced by the nervous system, and by the apprehension of mental phenomena. What was notable about Hope's *Treatise*, however, along with many other textbooks of the early nineteenth century, was that concepts of functional disease (that may or may not have any links to actual, structural changes detectable at autopsy), and of organic or structural disease (characterized by changes in the structure of the organ), were increasingly differentiated. Functional disease was associated with emotional neuroses, and with women. In the early part of the nineteenth century, functional disease did not have such moralistic connotations: functional disease could accompany structural disease, and its symptoms, palpitations, arrhythmia, heart pain, were classed as disease symptoms in their own right. However, Hope referred as early as 1832 to functional disorders as those that he had 'frequently met with in hysterical females subject to palpitation and in cases of nervous dyspepsia and hypochondriasis, under the form of spasmodic, aching pains in the anterior part of the chest'.[63]

There were many other textbooks on cardiac disease published during the nineteenth century, amongst them Walter Hayle Walshe's *A Practical Treatise on Diseases of the Lungs and Heart*, a work originally published in 1851, which went into four editions, the fourth edition being published as *A Practical Treatise on the Diseases of the Heart and Great Vessels, Including and Principles of their Physical Dagnosis* in 1873.[64] Most works dealt extensively with rheumatic fever and other perceived structural diseases, initially sidestepping the problem of functional diseases or emotional impact. By the later nineteenth century, textbooks dealt with the problem by accommodating emotional impact, first under the category of 'functional diseases of the heart', and subsequently as a separate category, marked 'cardiac neuroses'.

An important example, written as a textbook by a practising physician, was *Diseases on the Heart and Thoracic Aorta*. Its author, Sir Byrom Bramwell, was a physician and neurologist born in Northumberland.[65] Bramwell studied medicine at Edinburgh University before becoming house surgeon to James Spence, Professor of Surgery. In 1869 he returned to his father's family practice, and was appointed lecturer in medical jurisprudence and pathology at Durham University. In 1874 he moved to Newcastle upon Tyne and took up a post as the physician and pathologist at Newcastle Royal Infirmary. A move back to Edinburgh followed in 1879, where Bramwell became a fellow of the Royal College of Physicians.

In addition to lecturing on medicine in Edinburgh, Bramwell became pathologist at the Edinburgh Royal Infirmary, and was made a physician in 1897. He remained in this post until 1912. He also published extensively in the field of medical practice, writing nine volumes of clinical reports, ten books, an *Atlas of Clinical Medicine*, and over 160 papers. He specialized in the diseases of the spinal cord, blood, and aphasia. *Diseases of the Heart and Thoracic Aorta* was published in 1884, and originally delivered as lectures to the author's Edinburgh class in 1883–4.

In this work, Bramwell included 'diseases of the heart and pericardium' as diseases of the circulatory system, alongside 'diseases of the arteries' and 'diseases of the brain'.[66] The heart was described as a 'muscular pump', or 'two muscular pumps—the systemic and pulmonary respectively'.[67] The structure and operation of the heart were diagrammatically delineated. Referring to Michael Foster's *Textbook of Anatomy*, however, a work Bramwell described as 'the first English textbook of wide influence', he acknowledged the need to think of the heart 'both as a mechanical pump and as a vital organ' in order to understand its physiology and pathology, and the symptomatology, diagnosis, and treatment of its diseases.[68] Unlike others, therefore (including the physiologist Walter Holbrook Gaskell), Bramwell did not think that the operation of the heart was grounded in its material structure, but in a 'rhythmical property possessed by the muscular tissue independently of a special nervous mechanism'.[69] And, although the movements of the heart were automatic, the organ was 'intimately connected' with both the sympathetic and cerebrospinal nerve centres, and its action can, as each one of us so well knows, be readily affected by general 'nervous influences'.

Those nervous influences could be understood, he continued, only by recognizing the complex nervous supply of the heart.

For Bramwell, then, it was the suspension of the heart in a network of sensitive nerves that made its relationship with the mind explicable. The sensations of emotions were transmitted, according to a newly emergent neurophysiology, through the nervous system and to the heart. The 'surface and substance of the heart' itself were etched with 'delicate nerve fibres'.[70] There was a recognized cardiac plexus 'junction' through which impulses passing to and from the heart were transferred from one nerve path to another, and 'by means of which communications are established between the heart and distant parts'. The 'main lines' of communication passing between the nerve centres and the heart were identified as the sympathetic and pneumogastric. Movements in the heart were believed to be automatic, 'i.e. are due to impulses arising in the heart itself', but were nevertheless 'profoundly modified by the condition of distant parts'. Those modifications were 'conveyed to the heart through certain branches' of those pneumogastric and sympathetic nerves.[71]

While he listed structural diseases as disorders of the circulation, Bramwell saw 'functional' diseases as those with 'no distinct morbid anatomy': 'no changes were found after death in the heart itself.'[72] This was in keeping with traditional attitudes towards functional disease—as was the assertion that functional diseases could be brief or long in duration. And yet there was an important distinction from earlier periods. Functional diseases were not, Bramwell noted, life threatening: they seldom if ever destroyed life, and were not typically followed by any permanent injurious effects. In some cases functional diseases could result in organic disease—in exophthalmic goitre, for example, 'palpitation and accelerated action of the heart are prominent symptoms' that could lead to hypertrophy and dilatation, but these were exceptional. In the main, they were due to 'derangement of the nervous mechanism of the heart; and the primary cause is very often located in some distant organ'[73]

Gendering Functional Disease

Bramwell's work emphasized the emotional or psychological nature of functional heart diseases to a degree not common in earlier decades. His work marks the shift taking place from functional heart diseases as recognized (if perplexing) actual diseases to neurotic and imagined products of the mind. 'In some cases', he notes, 'the primary lesion (if we may use the term in connection with hysteria, hypochondriasis etc.) is *cerebral*. Under this head are included the derangements of the heart which are so frequently met with in hysteria, hypochondriasis, and the like.' In other cases, it was the 'cervical portion of the *spinal cord*' that was 'at fault', producing palpitations and rapidity of the heart's action in cases of myelitis and locomotor ataxy. In a third group of cases, moreover, 'the primary lesion is situated in the cervical *sympathetic*; and in this group we are probably right in placing the derangements of the heart which are such striking symptoms in the affection termed exophthalmic goitre'. Fourthly, the derangement of the heart might be due to '*reflex* irritation, the primary cause being situated in some peripheral organ, such as the uterus or ovary'.[74]

The gendered nature of Bramwell's theory of neurotic heart defects is clear. It was women who were prone to those functional derangements, 'which accompany *hysteria* and *anaemia* and the cardiac affections which are met with in *chorea* and *exophthalmic goitre* are very much more common in the female sex'. In a footnote, Bramwell added that male diseases of the heart, by contrast, tended to be those 'which are due to atheroma, and gout, to strain, syphilis, and alcohol . . . aortic lesions, aneurisms, and true angina pectoris are examples in point'.[75]

These distinctions were also generational. Young people were subject to different diseases from the middle-aged and the elderly.

Leaving aside congenital diseases, which affected infants, 'functional affections of the heart' were most common in 'youth and early adulthood', which included ages from puberty up to the age of 30. Bramwell does not account for this tendency, though it is likely that the emotional excesses of youth were viewed as responsible for 'functional affections'. At the time of 'active manhood', lesions of the heart and arteries were most commonly due to strain, syphilis, and drink, with aneurisms and aortic valvular lesions appearing. And in 'later adulthood and old age', fatty degeneration, dilatation, and valvular lesions were believed to cause a large proportion of the deaths in men between the ages of 55 and 60. In women, the period between 45 and 50 seemed 'particularly critical'.[76]

The Canadian-born physician and cardiac specialist William Osler endorsed most of Bramwell's analyses. Like Bramwell, Osler studied the heart from the perspective of a morbid anatomist, and he classified its diseases in a similar manner. In 1892, Osler published *The Principles and Practice of Medicine*, a work that became the most widely used medical textbook of the early twentieth century.[77] Osler's description of diseases of the heart initially covered seventy-two pages and six categories, 'the odd one out', according to Christopher Lawrence, being neuroses of the heart, 'which covered ten pages and comprised palpitation, arrhythmia, tachycardia (an abnormally fast heart rate), bradycardia (an abnormally slow heart rate), and angina pectoris'.[78]

With the exception of cardiac neuroses, a category that had expanded to cover many earlier and respected forms of cardiac dysfunction, heart-disease cases were redefined as structural entities, each of which possessed its own natural history and symptomology.[79] The required skills for diagnosis on the part of the physician included palpation, percussion, and auscultation of the chest, and analysis of the volume and strength of the pulse, rather than its rhythm. These were the new 'cornerstones of diagnostic practice'.[80]

By the time of the third edition of *The Principles*, published in 1898, Osler referred to the changes that had taken place since 1892, each of which had influenced the revised structure of his textbook. Changes had been made to various sections, including cancers and gastric neuroses, and new sections were added to the 'diseases of the blood, heart, lungs and kidneys'.[81] His original section on 'Diseases of the Nervous system' was also rearranged, reflecting his attempt to 'group the diseases in accordance with the modern conceptions of anatomy and functions of the parts'.[82] Listing heart diseases as diseases 'of the circulatory system', like Bramwell, Osler devoted considerable time to symptoms, signs, and diagnosis of organic or structural diseases. The only form of structural diseases that was linked to emotional distress was pericarditis, which could cause 'great restlessness' and 'insomnia' before the patient became 'melancholic', and possibly exhibited 'suicidal tendencies'.[83]

The newer additions to Osler's *Principles* included sections on morbid anatomy and on bacteriology. In Section V, we find the revised subject of 'Neuroses of the Heart'—under which heading he placed palpitation, a symptom that was particularly common in those who suffered from 'increased excitability of the nervous system'. As in Bramwell's discussion, nervous disorders of the heart were gendered:

Palpitation may be a marked feature at the time of puberty, at the climacteric, and occasionally during menstruation. It is a very common symptom in hysteria and neurasthenia, particularly in the form of the latter which is associated with dyspepsia. Emotions, such as fright, are common causes of palpitation. It may occur as a sequence of the acute fevers. Females are more liable to the affection than males.[84]

While some forms of palpitation were produced by stimulants such as tea, coffee, and alcohol, and in rare cases of structural diseases of the heart, it was most commonly a 'purely nervous phenomenon—seldom associated with organic disease'.[85] In many cases 'violent action' and 'extreme irregularity' characterized the actions of the heart, although there need be no consciousness of disturbance. 'The irritable heart described by Da Costa,' Osler continued, 'which was so common among the young soldiers during the civil war, is a neurosis of this kind'.[86] Like palpitations, tachycardia (a quickened heartbeat) and arrhythmia (an irregular heartbeat) were usually due to 'cen-tral—cerebral—causes; either organic disease, as in haemorrhage, or concussion'; more commonly 'psychical influences'.[87] By referring to Da Costa, Osler was describing a problem that grew more common after the American Civil War.

The most extensive work to have been written on functional heart disease and definitions for this period is Charles F. Wooley's *The Irritable Heart of Soldiers*.[88] The historical problem of the 'soldier's heart' was a phenomenon originally described by Jacob Mendes Da

Costa after the American Civil War, in which combatants experienced chest pain, palpitations, shortness of breath, and tachycardia.[89] The physician Sir Clifford Allbutt—best known for his *System of Medicine* in eight volumes—maintained that this was a structural disease (he also insisted that angina pectoris was structural, and of aortic origin).[90] By contrast, the physician Sir Thomas Lewis concluded, while working at the Military Heart Hospital in Hampstead, London, that 'soldier's heart' was a functional or neurotic disorder that should be renamed 'effort syndrome'.[91]

That Lewis's interpretation gained most support is indicated by medical reworking of 'effort syndrome' as 'neurocirculatory asthenia', a condition that is believed to be of psychological origin. This provides an interesting twist on the gendering of functional disease. It is likely that soldiers afflicted by neurotic heart disorders were feminized, or emasculated, by this association, particularly in the context of the gendered and heroic construction of the man as soldier.[92] In the present day, 'soldier's heart' has been retrospectively reinterpreted as a form of post-traumatic stress disorder, a reclassification of emotional distress that is similarly political and contentious.[93]

Contemporary debates over the 'soldier's heart' touched on contemporary physiological investigations into the control of the heartbeat and cardiac muscle contractility. It also drew from, and influenced, some of the most important historical debates about how far the emotions could impact upon the heart's structure and function. This recategorization of the 'soldier's heart' did not, of course, solve the problem of the apparently vital hold over the heart that was held by emotions. Nor was there any easy correlation between scientific theorizing on functional heart defects and medical practice outside the military—this is a theme taken up in the following chapter in a case study of Thomas Arnold, a much-discussed Victorian educator (and a less-discussed victim of angina pectoris).

Again like Bramwell, then, Osler had delineated several forms of functional disease under the term 'neurotic' or 'nervous' disease. One important example was his treatment of angina pectoris, a disease that was originally identified as a structural disorder with changes in the heart detectable at autopsy. In a section devoted to 'False angina', Osler identified two main groups: 'the neurotic and the toxic'. The latter of these groups included the abuse of stimulants such as caffeine and nicotine, known to alter the course of the heartbeat. The former embraced 'hysterical and neurasthenic cases, which are very common in women'.[94]

Osler devoted large sections of *The Principles* to hysteria and neurasthenia, both of which were believed to impact negatively on the operations of the heart by producing functional disease symptoms. Hysteria was particularly common in women, and described as a 'state in which ideas control the body and produce morbid changes in its functions'.[95] A woman's predisposition to hysteria could be both hereditary and environmental—failure to inculcate correct habits of emotional 'control' at a young age, for instance—or even constitutional, the disease being located, as its name and etymology suggests, in the reproductive organs of women.[96] The more immediate and 'direct' causes, however, were 'emotions of various kinds, fright occasionally, more frequently love affairs, grief and domestic worries'. One of the 'visceral manifestations' of such troubles was 'cardio-vascular':

Rapid action of the heart on the slightest emotion, with or without the subjective sensation of palpitation, is often a source of great distress. A slow pulse is less frequent. Pains about the heart may stimulate angina, the so-called hysterical or pseudo-angina, which has already been considered. Flushes in various parts are among the most common symptoms. Sweating occasionally occurs.[97]

Neurasthenia was described by Osler as 'a condition of weakness or exhaustion of the nervous system, giving rise to various forms of mental and bodily inefficiency':

The term [neurasthenia], an old one, but first popularized by Beard, covers an ill-defined, motley group of symptoms, which may be either general and the expression of derangement of the entire system, or local, limited to certain organs; hence the terms cerebral, spinal, cardiac, and gastric neurasthenia.[98]

The treatment for neurotic or functional heart complaints was typically to try to reduce the patient's emotional distress, or 'to get the patient's mind quieted'. Osler also recommended moderate exercise, the keeping of regular hours, tepid, but not hot or 'Turkish' baths, and no stimulating drinks or concoctions. Rich foods and sexual excitement should both be resisted. Medications include iron, strychnia, aconite, *veratrum viride* (a herb also known as False Hellebore or Tickleweed), and, in 'obstinate cases', digitalis.[99]

In contrast with the newly emerging concept of neurotic disorders, which included emotional causation, organic diseases were separate and distinct for both Bramwell and Osler. Each had a 'distinct morbid anatomy', were 'acute and chronic', and 'primarily cardiac', meaning that they were specific to the organ and its 'muscular substance'. This designation included congenital malformation, mechanical rupture, 'growths', and 'inflammations', as well as the influence of other diseases, including rheumatic fever and scarlet fever.[100] Most symptoms of organic heart disease—dropsy, cough, shortness of breath—were related to 'derangement of the venous or arterial circulation of distant parts of organs', any symptoms experienced in the heart itself (such as palpitation), 'often insignificant'.[101] Organic diseases affected the circulation in two ways. First, they could impair the force of the cardiac muscle as a pump. Lesions, for instance, might weaken the cardiac muscle, impairing its 'driving' power and producing slow circulation. The second adverse affect was by the production of structural alterations in the 'valvular orifices and valve flaps' that interfered with the valvular mechanism and impeded the circulation.[102]

The 'New Cardiology': Experimental Physiology and the Heart of Science

The early decades of the twentieth century saw greater interplay between scientific innovations and medical practice, partly in identifying and classifying functional and structural diseases, but also in terms of the development of the 'new cardiology': a science of the heart that defined disease in terms more of 'disordered physiology' than of pathological anatomy or psychological neuroses.[103] The bridging of communication between clinical practice and experimentalism was evident in the activities of two individuals whose work has been referred to above: James Mackenzie (later Sir James), the general practitioner who conducted clinical research, and the scientist Thomas Lewis, who also practised clinical medicine.[104]

Mackenzie's work in particular marked the shift from studying the structure of the heart to studying its functions, a task he set about through traditional methods of physical examination, symptom analysis, and—his own invention—the use of a polygraph. In all this he was influenced by the animal experimentation being conducted by his contemporary, the physiologist Walter Gaskell.[105] With this emphasis on function and physiology came not only the gendering of functional diseases, but also a downgrading of the importance of correlation between clinical and post-mortem findings.[106] Yet what defined the technological enterprise that marked the heart of science from the mid-nineteenth century was, above all else, its scale.

As we have seen in previous chapters, there was nothing particularly new to the nineteenth century about the investigation of the heart as the site of life, emotions, or self-hood. What *was* new was the breadth of activity that marked out the heart as an organ, and that explored its meanings as a structure that was simultaneously vital and material, that was linked to the body and its processes as a pump for the movement of blood, and connected, too, with the mysterious processes of mind.[107] From the mid-nineteenth century, experimental physiology became 'intense and continuous', and was typically pursued within an institutional context.[108]

By the mid-nineteenth century, nervous control over the viscera, including the heartbeat, was a central concern for experimental physiologists. The heart's relationship to the sympathetic nerves had been asserted as far back as the seventeenth century, in the work of Thomas Willis and others.[109] Experimentation into the heart, and the vagus nerve in particular, had been conducted from the early nineteenth century, largely through the analysis of the rhythm of the heart.[110] From the 1860s, experimental physiology was impacted upon, and influenced, by the new measuring and objectification techniques that influenced the diagnostic categories of heart disease outlined above. One central issue that subsequently came under scrutiny was the origin of the heartbeat itself. With the development of experimental physiology and the decline of the soul as an explanatory system, constructions of the heartbeat as a phenomenon began to be reinterpreted.

Traditionally, the rhythm of the heart had been understood in two main ways: first, that the heartbeat was an intrinsic quality in the muscle of the heart, and, second, that it was produced by an extrinsic quality fed *into* the heart, whether through the nerves, the humours, or the spirits. By the eighteenth century these debates had resulted in debates over vitalism versus mechanism, and over the existence or otherwise of a human soul. In the nineteenth century, the sympathetic and parasympathetic nerves linking the heart were identified, and it became possible, and desirable, to trace nervous connections from the heart to the brain.[111] There were, however, problems. For how was it possible that the heart continued to beat even when excised from the body? Even within the body, experiments on the vagus nerve revealed some degree of cranial influence, but not the origination of the heartbeat.

In 1886, after a series of experiments into the 'involuntary' nature of the heartbeat, the British physiologist Walter H. Gaskell described two sets of nerves at work in the vascular and visceral systems, one of which was responsible for contraction, the other for relaxation.[112] It was Gaskell's Cambridge colleague John Newport Langley who introduced the terms 'ganglionic' and 'postganglionic' to describe these two nerve sets, subsequently giving the autonomic nervous system its name in 1898.[113] One of the questions that came out of these experiments and debates was the extent to which the heart and its functions were a product of the brain and *its* functions—a subject addressed more fully below.[114] This is interesting because it highlights broader concerns about emotions and the meaning of 'feeling', and the extent to which heart palpitations, for example, could be seen as products of mental and cognitive or bodily and automatic processes.

Otniel E. Dror has analysed the drive towards establishing categories for emotions and their representations in nineteenth- and twentieth-century scientific culture.[115] By the twentieth century, he argues, the laboratory had become 'an ecology of experiences and emotions' as researchers suggested new emotion-sensitive laboratory procedures; 'identified, recorded, and controlled emotions during various laboratory manipulations; devised new models for interactions between human and animals; and introduced qualitative emotional descriptors into the quantitative language of their experimental science'.[116] This experimental enterprise ran broadly parallel to the development of cardiology as a specialism, and of experimental and clinical attempts to understand the heart as an organ. Each of these initiatives meant that attitudes towards the heart continued to move towards a more comparable, precise, and quantitative standard. The quest for such objective measurement drew from and spurred on a series of cardiological monitoring devices that were less immediately relevant to bedside medical practice than to laboratory measurement of emotion states.

In 1859, for instance, the Parisian physician Jules Étienne Marey demonstrated his sphygmograph before the *Sociéteé de Biologie*. He published his findings in 1860.[117] This was not the first recorded attempt to develop the sphygmograph. An earlier design was devised by the German physician Karl Vierordt in 1854, though the quality of representations it produced was believed to be inferior.[118] The techniques of Marey's sphygmograph were rapidly emulated. Robert Frank has described its innovation as

merely the first in a series of instruments, such as the capillary electrometer, the polygraph, the electrocardiograph, the phonocardiograph, and the sphyg-momanometer, that were developed over the succeeding half century and that aimed to make the salient features of the action and pathology of the human heart accessible to the physiologist and the physician.[119]

Although the pulse had long provided a means to analyse the workings of the individual heart, technological innovations offered to describe the movements of the heart in a measurable and calculated way. This is 'an exquisitely designed instrument', claimed the editor of the *Lancet* in 1865, 'by the aid of which the pulse is armed with a pen, and at every beat writes its own diagram, and registers its own characters'.[120] Marey's cardiological invention was designed to produce a written graph of the movements of the pulse through the mechanical transmission of an impulse to a recorder. It was explicitly designed to produce an infallible and graphic representation in order to remedy 'the defectiveness of our senses for discovering truths, and secondly the inadequacies of language for expressing and transmitting the truths that we have acquired'.[121]

By the second half of the nineteenth century, the desire for precision in experimental physiology, especially that indicated through the application of the graphical method, had become crucial to the application and dissemination of scientific knowledge, and to the construction of 'emotion' as a quantifiable laboratory object.[122]

The Limitations of Practice

We might view these innovations as revolutionizing the clinical encounter, as the heart of science beginning to dictate medical practice. And yet this is too simplistic. One of the most interesting aspects of the technologizing of the heartbeat in Britain from the late nineteenth century is how often practising physicians spurned it. Christopher Lawrence has convincingly argued that, for many practitioners of the late nineteenth century and beyond, there was a resistance to the use of technological instruments in the bedside context. Although the use of the language of science was an important weapon in the professionalization of the medical profession, and of individual physicians' attempts to show their learning, the status of clinical medicine ultimately rested on its being something other than (something more than) a science. Thus, what Lawrence refers to as an 'epistemology of individuals' experience', of experience learned through the clinical encounter, was invoked by British physicians to defend the autonomy of British medicine.[123] Making knowledge 'incommunicable', in other words, helped provide the illusion of objectivity.

As late as 1905, therefore, the *British Medical Journal* could lament, in reference to the sphygmomanometer (invented by a French physician, Julius Herisson, in 1835 with the purpose of making the pulse visible by transmitting its beat to a column of mercury), that 'by such methods we pauperize our senses and weaken clinical acuity'.[124] By this time there were many mechanical instruments to aid physicians—including the sphygmomanometer, ophthalmoscope, sphygmograph, spirometer, laryngoscope, and microscope—but the British medical press, and individual physicians, continued to show a suspicion of instrumental diagnosis.[125] Hughes Evans notes a similar process of blood-pressure analysis more generally, with many clinicians disputing the relevance of new laboratory sciences to their practice.[126]

This reluctance to embrace technological innovation into clinical practice seems to have been peculiarly British, or perhaps English, in character.[127] On the Continent, especially in experimental physiology, technological representations of heartbeats (and emotional states more generally) were becoming much more popular. After conceiving of the sphygmograph, Marey invented additional graphic devices, including apex-cardiograms that were applied to the chest wall to measure cardiac movement. These proved popular; Marey's cardiograph was taken up by the French physiologist Claude Bernard (1813–78) in his measurement of cardiac movement in emotional states.[128] Bernard's influence on his pupils was such that he encouraged English physiologists to make use of its development; having studied in Paris under Bernard, John Scott Burdon-Sanderson was keen to introduce the sphygmograph into British physiological research.[129]

The heart's status in these developments was seldom problematized. Dror has noted that both Bernard, and his follower the Russian physiologist Elie de Cyon, unusually focused on the heart as both an organ of emotion *and* an organic structure: in their lectures 'they described the biological mechanisms that collapsed the heart's physiology (as a pump) with its psychology (as an organ of emotion)'.[130] Yet no sustained programme of research followed this attitude towards the heart as an organ *of emotion*. Nor were Bernard's and de Cyon's questions and measures influential in future physiological research. While the science of emotion was to become a brain-centred one—physiological instruments being used to measure what Angelo Mosso referred to as precise measurements of the 'physiological concomitants of emotions *and mental states*', the heart became a biological variable, to be studied alongside other such variables as blood pressure, body temperature, glucose level, adrenalin levels, and cell count.[131]

Twentieth-Century Developments

By the start of the twentieth century, the medical specialization that has been described by Rosen, and others, as in general application throughout British medicine was beginning to become evident in the cardiac field.[132] A growth of hospitals specializing in heart disease—such as the London Hospital for Diseases of the Heart and Lungs (1848) and the National Hospital for Diseases of the Heart (1857)—as well as the emergence of late-nineteenth-century cardiac specialization as an area in which up-and-coming medics might make their mark meant that the work of experimental physiologists began to feed back into institutionalized medicine, reshaping notions of the heart and producing what Christopher Lawrence has termed the concept of the 'living heart'.[133] The impetus was not solely a product of experimental physiology and technological change. Developments in other areas—including bacteriology, pharmacology, and even epidemiology—led to an interest in the heart to elucidate various aspects of health care and treatment. In the context of bacteriological developments, for instance, physicians studied the heart to understand acute and chronic valvular disease.[134]

The initial development of specialization within medicine was not one that was wholeheartedly embraced, perhaps a nod to physicians' desire to preserve the 'incommunicable knowledge' outlined above. Moreover, there was a genuine philosophical resistance to specialism on the part of many; thus, Burnley physician and cardiologist James Mackenzie could express with concern that 'the day may come when a heart specialist will no longer be a physician looking at the body as a whole but one with more and more complicated instruments working in a narrow and restricted area of the body'. And that, he concluded, 'was never my idea'.[135]

Experimental Physiology and the 'Science of Emotion'

By the early twentieth century, 'emotions' had become a legitimate end to physiological investigation, with researchers like Walter Bradford Cannon investigating emotions in various organ systems and functions.[136] While the rise of the mind sciences meant that the origin of emotions was most often rooted in the brain, the heartbeat was largely disassociated from emotions except as a symptom and evidence of emotional experience. In short, the heartbeat of experimental physiology could be interpreted—like sweating and blushing—as just another by-product of physiological processes controlled and regulated by the brain.

Between the eighteenth and nineteenth centuries, some profound changes took place in the understanding and articulation of heart disease, several of which have been touched on in this chapter. First, heart disease emerged as a clinical specialism, along with several subsectors of heart disease, including angina pectoris. Secondly, techniques of measurement and standardization moved the diagnosis of heart disease away from the subjective assessment of the patient to the objective assessment of the physician, and to the visual representation of pulse rate through such technologies as the sphygmograph and the electrocardiograph.

Thirdly, we have seen how new technologies and methods helped to shape notions of heart disease itself in the nineteenth century; the identification of new series of symptoms (related to the sounds in the chest, for instance) providing new ways by which cardiac problems could be measured. The use of the stethoscope helped to localize disease in the organs, and proved something of an impetus to the development of anatomo-clinical medicine.[137] At the level of theory, if not practice, technologies that focused on the function rather than the structure of the heart helped shift cardiology's development as a discipline towards the hearts of the living, and away from those of the dead. During that process, subjective experiences of emotion lost influence in favour of physicians' interpretations of structural and functional disorders.

Angina pectoris was a case in point. In John Hunter's day, the newly identified disorder straddled traditional and reworked ideas about emotions and their ability to cause physical or functional changes in the heart. By the mid-nineteenth century, the divisions between structural and functional diseases became more rigid, and more gendered. At the same time as the division between functional and structural disorders became clearer, functional disorders were feminized, while structural diseases were reinforced as masculine conditions. This gendering of heart disease created an association between middle-aged men and physical diseases of the heart that remains in place in the early twenty-first century.

4

Angina Pectoris and the Arnold Family

I heard a rattling in the throat and a convulsive struggle. I called out, and turning to him I supported his head, which was thrown back, on my shoulder. His eyes were fixed and his teeth set, and he was insensible. His breathing was very laborious, his chest heaved and there was a severe struggle over the upper part of the body. His pulse was imperceptible, and after deep breathings at a few prolonged intervals all was over. He died in little more than half an hour after I first saw him.

Peter Mere Latham, *Lectures* (1845).[1]

The Case of Thomas Arnold of Rugby

On Saturday, 11 June 1842, Thomas Arnold, educational reformer and Headmaster of Rugby School,[2] strolled in the institutional gardens with his head boy, William Charles Lake.[3] Thomas would subsequently dine with the sixth form boys—a customary act on the last day of term. His diary entry for that evening noted that 'the day after to-morrow is my birthday, if I am permitted to see it'.[4] His words proved prescient. At 5 a.m. on Sunday, Thomas woke with agonizing pains across his chest. Three hours later he was dead, a victim of angina pectoris. The same disease had killed Thomas's father, William Arnold, the Collector of Customs and Postmaster of the Isle of Wight.[5] It would also kill Thomas's son, the poet Matthew Arnold. Each of these three generations of Arnold men diagnosed with angina pectoris died suddenly and unexpectedly: both Thomas Arnold and William Arnold were in their forties; Matthew was 65.

Despite the historical importance of Thomas and Matthew Arnold as literary, educational, and political figures, there has been little attention given to their apparently congenital heart complaints. Rather more interest has been paid to the leg callipers that Matthew Arnold wore to combat rickets.[6] And yet, like the case of the eighteenth-century surgeon John Hunter, that of Thomas Arnold opens up something of a case study by which we can explore the meanings of heart disease in a particular social context, and at a time when attitudes towards heart disease were apparently changing.

Through the case of Thomas Arnold this chapter engages with several important critiques on the nature of heart disease as a concept in Victorian Britain. These include Kirstie Blair's analysis of the poetics of the heart as both organ and symbol in Victorian literature; Christopher Lawrence's path-breaking descriptions of the emergence of cardiology as a medical specialism; and Charles F. Wooley's detailed account of the emergence of 'nervous-heart' syndrome, identified within the US military.[7] Despite the considerable merits of these accounts, what is missing is what a *longue durée* approach to the history of emotions and the heart can bring: a systematic analysis of change and continuity in the relationship between emotions and angina pectoris.

This chapter has three main aims. First, it will demonstrate the continued centrality of emotions to the heart and its diseases in nineteenth-century medical discourse. As in the eighteenth century, this centrality was largely a result of long-established interconnections between *psyche* and *soma*, and yet it had not been displayed by the rethinking of the mind-body relationship that had taken place in the eighteenth century. This is not simply to say that things did not change, or that there was more consensus than conflict. Rather, it is to suggest that we need to acknowledge this shifting yet stable legacy in order to be able to appreciate and understand one of the most significant aspects of disease concepts that emerged during the nineteenth century: the problematic nature and relationship of organic and functional disease entities. These depended on broader interpretations of the mind-body relation, and emotions became crucial components in the conception of cardiac symptoms as 'functional' or 'structural' (or organic) in origin. To clarify, the term 'functional' is used throughout this chapter in the sense employed by Wooley, to 'describe disorders of function of an organ or system without an obvious physical lesion'.

By contrast, 'structural' heart disease refers to disorders that correlated with structural changes identified at autopsy.[8]

As we saw in the case of John Hunter, emotions were traditionally believed to impact on the body at both a structural level—by contributing to arterial deterioration, for example—and a functional level—by causing palpitations. This reinforced the considerable, often spiritual, weight given to the emotions as conduits between mind and body, spirit and matter. And yet there were distinct, largely unacknowledged, changes taking place in attitudes towards structural and functional diseases during the nineteenth century. These drew on changing perspectives of the mind-body relationship, and the respective role of the heart and the brain in localizing emotional cognition and experience.

A connected claim made in this chapter is that historiographical accounts of structural and functional disease tend to be simplistic in their neglect of these changes, largely because they derive from a modernist perspective and—unlike a *longue durée* approach—cannot address shifting patterns of emotion rhetoric and theory over time.[9] The history of angina pectoris does not readily conform to the model of heart disease identified by Wooley, in which (to put it crudely) diseases were identified as functional until such time as sufficient diagnostic tools existed to label them as structural.[10] Throughout the nineteenth century there was much flexibility and conflict over the relationship between psychological and physical causation, and between the symptoms of structural and functional disease. The two were never mutually exclusive or independent. It is simply incorrect, therefore, to claim that 'functional' was 'the opposite of "organic" disease' in nineteenth-century medicine.[11]

Finally, this chapter follows Charles Rosenberg, John Harley Warner, and others in emphasizing the difference between medical therapeutics, or practice, and medical theory.[12] For, when we come to look at the treatment of Thomas Arnold, and the recommendations of Peter Mere Latham, the physician for whom 'T.A.' became a case study, we find that the heart was regarded in neither mechanical nor even material terms (see Fig. 5). Nor was it subject to the advances in pharmacology and technology that we have come to associate with the 'scientific' and rational nineteenth-century medical tradition.[13] Instead, we find the continuation and preservation of eighteenth-century medical practices.

Fig. 5. Peter Mere Latham. Stipple engraving.
Source: Wellcome Images.

Along with many others of his generation, Latham maintained a holistic interpretation of cardiac disease that emphasized links between the heart, the emotions, the blood, and the entire circulatory system. In exploring these areas, this chapter will begin with an examination of the status of heart disease in nineteenth-century culture, and its perceived increase as a result of the stresses of modern life. It will also scrutinize further nineteenth-century debates on structural and functional causation, and explore the therapeutic practices governing heart health, mind, and body in nineteenth-century medical culture.

A Disease of Civilization?

The death of Thomas Arnold was unusually well publicized in nineteenth-century Britain—by Peter Mere Latham in his work on heart disease, and by Arthur Penrhyn Stanley, a student of Thomas Arnold who later became dean of Westminster. Stanley was commissioned by Mary Arnold to write her dead husband's biography.[14] In Latham's account, Thomas is termed 'T.A.' and discussed as one of three anonymous individuals who died of angina pectoris. Latham's work includes a detailed transcription of Thomas's last conversation with his attending physician, Dr Bucknill junior.

When examined, Thomas told Dr Bucknill that his own father had died of heart disease, and, apparently aware of the severity of his condition, asked whether this was usually fatal. When Bucknill replied in the affirmative, Thomas asked whether heart disease had become 'a common disease'. Bucknill replied that it was not very common, but increasingly 'prevalent in large towns, perhaps from anxiety and eager competition among the higher and intemperance among the lower classes'.[15] Bucknill's judgement about the increasing incidence of heart disease (particularly among urban populations) was consistent with intensified medical and popular interest in the study of the heart and its diseases.

Since its inception in the eighteenth century, and its fame as the disease that killed John Hunter, angina pectoris had grown in significance and reputation as a cause of death. As will be demonstrated in more detail below, it is apparent that Thomas Arnold and his wife Mary, as well as their family physician, were well aware of the existence, the symptomology, and the received wisdom over the causation and prognosis of angina pectoris. Between the late eighteenth and mid-nineteenth centuries, however, diagnoses of heart disease in general had undergone a profound change, both in terms of the categorization and construction of disease concepts, and in the role of the heart as an organ of disease. Interestingly, however, the link between heart disease and emotion remained intact.

As indicated by Bucknill's comments, there was a growing belief in nineteenth-century Britain that heart disease was on the increase. This belief encouraged, and grew from, a relatively new growth in cardiac medicine as a clinical specialism, with the emergence of specifically qualified physicians and hospitals, and an apparent rise in heart disease as a cause of death.[16] In 1872 *The Times* reported that male deaths from heart disease had risen from 5,746 in 1851 to 12,428 in 1870. This shift was linked to the 'working years of active social life' and particularly amongst men aged 21–40 years.[17] In an interesting parallel with modern medical practice, nineteenth-century heart disease was, therefore, gendered towards men. This is consistent with the earliest diagnosis of angina pectoris as a symptom complex.[18] Part of this increase in heart disease, the *Times* article continued, resulted from the 'great mental strain and hurried excitement of these times, in an overcrowded community, where competition is carried to the highest point'.[19]

This was not the first time that an increase in heart disease had been linked to 'mental strain and excitement'. In 1809 Jean Nicolas Corvisart, personal physician to Napoleon Bonaparte, claimed that there were more deaths 'from the organic affections of the heart' than from 'those of the lesion of the brain, stomach, liver, spleen, kidneys &c. taken together'. Part of the reason, Corvisart claimed, was the fall-out of the French Revolution, a period that saw unprecedented levels of national and individual anxiety. There were, therefore, in other words, two reasons for a contemporary growth in diseases of the heart: 'the action of the organ' and 'the passions of men'.[20]

The psychological state of Thomas Arnold prior to his death is regrettably unknown. We do not know whether he expressed anxiety about his health prior to the day before his fatal attack, when he experienced some chest pain after bathing.[21] Thomas did acquire during his lifetime a reputation for unrelentingly hard mental work, however, and at least one historian has suggested that his daughter's deep depression when her engagement ended contributed to Thomas's subsequent heart problems.[22] The association of emotional strain with heart disease, particularly in chronic states like 'stress', has resonance in modern interpretations.[23] In common with many other diseases—including cancer—heart disease has become a 'disease of civilization', a product (as suggested in the *Times* article cited above) of the peculiar demands of modern life.[24]

There is clearly no doubt that in the nineteenth century emotions continued to be regarded as important to heart health. Indeed, emotional extremes remained one of the most influential factors in the development of angina pectoris, as it had been since the disease's eighteenth-century origins (though its relevance was increasingly gendered). The disease related to exertion (both physical and mental), and to diet and health, all of which were traditional lifestyle factors to be monitored. Patients were typically struck down after eating or taking exercise, and the disease could develop over several years or—as in the case of Thomas Arnold—be immediately fatal.

Angina pectoris was also increasingly well recognized outside medical circles, evidently a result of its ubiquity in medical journals and discussions amongst the learned elite. When Thomas clutched his chest in the early hours of a Sunday morning, his wife Mary was immediately vigilant. Although Thomas had previously reported no heart problems to her, she was aware that his father had died young of heart disease. And when her husband told Mary that his pain 'seemed to pass from his chest to his left arm', it was reported that Mary's 'allarm [sic] was so much roused from a remembrance of having heard of this in connection with Angina Pectoris, and its fatal consequences', that she insisted Dr Bucknill be summoned at once.[25]

In addition to the well-established symptomology of angina pectoris, beliefs about the origin of the disease were originally straightforward. Angina pectoris had developed as a structural disease, prompted by physical or mental exertion that caused physical damage to the heart, particularly the coronary arteries. This claim is directly at odds with Wooley's analysis of angina pectoris as a functional disease, that only gradually, and over the course of the nineteenth and twentieth centuries, became redefined as a structural disorder. In Wooley's discussion, this is one of many cardiovascular, gastrointestinal, and neurological disorders to be 'defined, classified, and then redefined in each succeeding period of clinical medicine' from the seventeenth through the nineteenth centuries.[26]

Although originating as a functional disease, Wooley suggests, angina pectoris was only gradually redefined as a structural disorder by technological discoveries—including the measurement of arterial hypertension and cardiac arrhythmias—that underpinned medicine's development as a scientific discipline.[27] This perspective is compatible with Wooley's broader interpretation of the emergence of modern medicine and the invention of new diseases but perhaps misunderstands the nature of angina pectoris prior to the mid-nineteenth century, and the complexities of heart-disease concepts thereafter.

First, angina pectoris was always associated with structural disease, as seen in the case of John Hunter. Moreover, the likelihood of emotional extremes causing damage to the heart was increased in cases of pre-existing structural weakness. As the Glaswegian anatomist Allan Burns wrote in his *Observations* (1808), when the 'nutrient arteries of the heart [were] diseased', it took very little for the heart to become 'overpowered with blood accumulated in its cavities'. Then the pulse would begin to falter, the right ventricle would became unable to supply blood into the pulmonary vessels, and 'a sense of suffocation' would strike the sufferer. He would then feel 'an indescribable anxiety and oppression' as a result of accumulated blood around the chest.[28]

Rather than viewing angina pectoris as transforming from a functional to a structural disorder, then, we can see a growing desire and readiness on the part of nineteenth-century theorists to label diseases as *either* functional *or* organic, a product of wider shifts in the mind–body relationship and the rise of the mind sciences. This readiness was legitimized and facilitated by such innovations in diagnostic analysis as the stethoscope and sphygmograph. As Wooley acknowledges, the signs of organic and inorganic disease became 'much more precise and simple, as a consequence of the new lights thrown on particular valvular diagnosis and on inorganic murmurs'.[29] A converse way of looking at this phenomenon is that the emergence of more systematic and precise forms of measurement gave rise to a greater desire for standardization, and a correspondingly greater emergence of 'abnormalities' or 'disorders'.

The working-out of angina pectoris as a functional/structural disorder in the nineteenth century, therefore, can be viewed as part of the process of scientific rationalization about the heart as an organ. Debates over functionalism and structuralism continued throughout the nineteenth century before solidifying along gendered lines. Angina pectoris was increasingly included in treatises on the heart under such

headings as 'functional diseases', 'nervous disorders'. or the increasingly common term 'cardiac neuroses'.[30] This shift in the language
of diagnosis, and in the status of angina pectoris, is evidence of the
increasing tendency of medical theorists by the end of the nineteenth
century to regard most experiences of angina pectoris as, first, functional in origin, and, second, symptomatic, not of any underlying
structural disorder, but of some emotional disturbance or neurosis.
In the mid-nineteenth century, and at the level of medical practice,
however, definitions of angina pectoris, and of functional or structural
diseases, remained in flux. To consider this in more detail we need to
return to the subject of our case study, and to the body of Thomas
Arnold.

The Body of Thomas Arnold

Thomas Arnold's body was autopsied within forty-eight hours of his
death, 'the weather being very hot'.[31] Along with his final conversation
with his physician, Thomas's autopsy was reported in Latham's *Lectures*.
The autopsy was conducted by Joseph Hodgson, surgeon at the
Birmingham General Hospital from 1822, and author of *Treatise on
the Diseases of Arteries and Veins* (1815), and was witnessed by the
junior and senior Drs Bucknill.[32] On examination, Hodgson found
that the pericardium was healthy, but that the heart was 'rather large'
and 'very flaccid and flat in its appearance'.[33] There was no indication
of valvular disease, and the membranes were healthy. The muscular
structure of the heart, however, was 'remarkably thin, soft and loose
in its structure'. The walls of the left ventricle were also 'much thinner
and softer than usual'.[34]

Along with the other two cases reported by Latham, Thomas's
autopsy showed that structural changes had taken place in the heart,
the 'muscular substance' being 'reduced to an extreme degree of
tenuity and softness'.[35] It was clear to Latham that these changes had
contributed to the angina in each case. Yet Latham did not believe that
evidence of structural change in the heart denoted the organic nature
of angina pectoris. He acknowledged that angina pectoris could
exist where there was 'ossification or obstruction of the coronary
arteries', 'dilatation of the aorta', 'valvular unsoundness', 'hypertrophy
or atrophy', or a softening of the heart's muscular substance, as in

the case of Thomas Arnold.[36] Yet Latham maintained that angina pectoris had also been reported where there was 'no form of disease or disorganization' either in the heart or in its blood vessels.[37] Moreover, the pain involved in angina pectoris could be misleading, for dangerous heart complaints were not always painful. What this meant for Latham was that angina pectoris was nothing more than a symptom complex: 'an assemblage of symptoms' that had been 'made to bear the name of a disease'.[38]

Although angina pectoris was historically identified as an organic disease, its symptomology a result of structural changes that were associated with extreme emotions and lifestyle conditions, including diet and overall health, Latham offered a functional interpretation. Put simply, if gnomically, the 'paroxysm of angina pectoris is plainly a compound of pain and something else'. The heart ceases to function, perhaps because of a spasm that is in itself the whole disease: 'a disease purely vital, a disease of feeling and function alone, operating by and through a sound structure, it may be fatally, always perilously'.[39]

Unquestionably, Latham acknowledged, the symptoms associated with angina pectoris—the crushing pain in the breast and down one arm, the breathlessness and sense of panic—were widely characteristic, and occurred with sufficient regularity to be identifiable as angina pectoris. Yet, he claimed, they did not amount to any single unity or disease. There were differences in the structural changes identified in the hearts of his three case studies. And other cases had existed without structural alterations being detected. Although angina pectoris might be independent of organic disease (or produced by it), there was no easy correlation between the two.[40] Angina pectoris was a spasmodic attack that could be linked to structural change, but was not detachable from the general systemic health of an individual.

To avoid angina pectoris, and to some extent *all* forms of heart disease, Latham advised, one needed to take care of the 'whole vascular system'.[41] For the 'spasm' that had for so long been identified as a disease could be 'putt off' and its 'severity mitigated' by 'no means more surely than by keeping the vascular system in a just balance between fulness and emptiness, between rich blood and poor blood'.[42] Of course, there would always be those individuals who—in similar tones to humoral physiology—possessed a natural balance with which they were 'very happily born', but for most 'a small habitual deviation on this side or that is readily felt and resented by the heart, when it has undergone some form of unsoundness rendering it obnoxious to spasm'.[43]

What, then, are the implications of Latham's analysis for our historical understanding of angina pectoris and the case of Thomas Arnold? Much of the identification of nineteenth-century scientific medical culture presupposes the decline of holistic, system-based medicine in favour of 'organ-based' specialisms.[44] And yet Latham, one of the acknowledged fathers of modern cardiology, identified the heart as embedded within the entire mental and bodily system.[45] In the same spirit and with the same precision that eighteenth-century physicians recommended a strict adherence to the non-naturals, Latham advised patients to monitor their general lifestyles, to monitor the kind of food and drink taken, ensure sufficient sleep, avoid 'troublesome wants such as frequent micturation' [urination], and use only moderate physical exertion.[46] Finally, the 'passions and affections of the mind' must be regulated, for more than any other cause they produce fatal attacks of the paroxysm. A similar course of action was taken by Thomas's son Matthew Arnold in avoiding the onset of angina pectoris. In a letter to his son Richard Penrose, Matthew complained in 1885 that his physician, Andrew Clark, had put him on the 'strictest of diets for one week—no medicine, but soup, sweet things, fruit, and worst of all, all green vegetables entirely forbidden, and my liquors confined to one small half-glass of brandy with cold water, at dinner'.[47]

What is remarkable is that, despite the changes taking place in constructions of the heart and its diseases between the death of John Hunter and that of Thomas Arnold, little had changed in its interpretation, or in its therapeutics. In his treatment of Thomas Arnold, the physician Bucknill followed traditional prescriptives for the improvement of circulation and heart function. First, Bucknill ordered that Thomas be given hot strong brandy and water. He also arranged for a mustard plaster to be placed on Thomas's chest. Traditionally, mustard plasters were applied to stimulate the internal organs and assist respiration. While the mustard plaster was prepared, Bucknill himself applied hot flannels to Thomas's chest, while the patient's legs and arms were 'rubbed' by an assistant. Thomas's feet were then tightly wrapped in flannels that had been soaked in the hot water and mustard mixture. The treatment seemed to be effective, as Thomas's pulse became 'natural', his arms and legs were warmer, and Thomas was temporarily free from pain.

There is no evidence that Bucknill believed his attention would *cure* Thomas's disease, or do any more than temporarily alleviate his symptoms. The published record of this encounter suggests a pessimistic tone to the proceedings, with Bucknill acknowledging that painful spasms often returned as swiftly as they had departed, and that such symptoms were usually fatal. What is relevant here, however, is that the therapeutic strategies undertaken by Bucknill, and recommended elsewhere by Latham and others, were primarily aimed at soothing the patient, at increasing his temperature, and at encouraging the circulation of the blood. The pharmacological remedies used by Bucknill were also traditional—first, laudanum (an opiate) to relieve Thomas's physical pain, and camphor, often used as a local anaesthetic. More significantly, perhaps, Bucknill also turned to Hoffman's anodyne, essentially the spirit of ether, which was widely used to stimulate a weak heart after Friedrich Hoffman's hydro-dynamical investigations in the eighteenth century.[48] Digitalis, derived from the foxglove and widely believed to be used as a remedy for heart complaints by the early nineteenth century, was not used.

As Christopher Lawrence and George Weisz have recently argued, the survival of such medical holism challenges our preconceived ideas about the predominance of 'scientific' rationalism in nineteenth-century medicine.[49] It illuminates the long tradition of medical writings about emotional influence on the heart as an organ in the modern West, when classical accounts of mind-body relation are mainly metaphorical.[50] Yet it perhaps also has relevance for our modern understandings of the heart and the emotions. In the twenty-first century it is well established that cardiac dysfunction can be a product of emotional disturbances, whether long term, in diagnoses of stress and anxiety, or short term, as a result of shock. The challenge for historians must be to explore these apparent similarities in diagnostic categories, without homogenizing, or ahistoricizing, concepts and understandings of the heart both as figuratively embodied, in relationships of gender, class, and differences, and as a literally embodied organ.

Latham's use of Thomas Arnold as a case study tells us much about the therapeutic strategies used by practitioners in mid-nineteenth-century Britain. Less emphasis was placed on the measurement and objectification of cardiac symptoms than on traditional remedies to increase the circulation and relieve pain: mustard plasters, hot flannels, and opiates. To assess how far the preservation of traditional practices was common during the period we need to delve further into the records of practising physicians and emergent specialists like 'Heart Latham'. This is, indeed, the subject of the following chapter.

5

'Heart Latham' and Nineteenth-Century Medical Practice

Arrived at 3 in the morning and found a man pale with cold extremities and clothes steeped in ice water around the front of his chest and a bladder of ice lying on his sternum. A very feeble pulse—forbid to utter a word—and writing all he had to communicate on a slate.

Peter Mere Latham, *Medical Case Book* (1840)[1]

In the previous chapter, we saw how influential Peter Mere Latham had been in popularizing awareness of the death of the education-alist Thomas Arnold, and in identifying the main causes of angina pectoris in mid-nineteenth-century Britain. This chapter examines the work of Latham in more detail, with reference to the medical context in which he practised, and to his own casebooks, in which the above entry was made on 11 November 1840.

Latham kept detailed notes on each of the patients he had visited, recording their background and symptoms, and his diagnosis, as well as any prognostic developments. On this particular occasion, Latham described how he had been asked to visit Ventnor on the Isle of Wight with his colleague Dr George Anne Martin. The patient, John Forbes, was a 39-year-old male living in Reading.[2]

Latham recorded that, prior to his arrival at Ventnor, Forbes had been 'feverish', experiencing 'perspiration' and attacks of coughing that brought him 'sometimes to dyspnoea' or fainting. At his own request, the patient had been taken for recuperation to Ventnor, a small village on the Isle of Wight well associated with tuberculosis and cardiac care.[3]

Almost immediately after Forbes's arrival in Ventnor, he began to spit blood. At least 2 oz were collected by his caregivers, and on two separate days. The blood was 'dark coloured and unmixed' and without any other 'expectoration'. His treatment consisted in bleeding—he was bled twice from the arm—and purging. To ease his cardiac symptoms, he was given digitalis, a relatively new form of treatment. Digitalis had been mentioned in the *Lancet* from the 1830s, but, even a century later, its efficacy in cardiac cases was still disputed.[4] The cold-water treatment Forbes was receiving was, however, a standard practice in contemporary spa treatment.[5]

On Latham's arrival, the bladder of ice lying on the patient's chest was removed so that he might be laid down and examined. Latham used percussion to listen to his chest, and was able to note that there was 'no unnatural sound anywhere'. Nor was there any 'respiratory murmur' detected by the use of the stethoscope.

Latham watched Forbes over the next day and night and noticed that, when he had been taken out of his ice-cold and dripping clothes, the man's pulse 'rose a little'. He personally bled his charge, taking 'with much labour about four ounces from beneath the clavicles', before applying 'a large blister in front of the chest'. By the afternoon, Forbes reportedly became 'hot and a little flushed and his pulse was fuller'. He coughed up a large globule of mucus 'deeply stained with black blood', after complaining of discomfort in his chest. Latham left while the blister was still in place. He later received a letter from Dr Martin, informing him that the patient had subsequently excreted a copious amount of 'pitch-like matter' in his stools, for which Latham recommended purgatives that included calomel, a mercury compound that was 'to be pushed until the evacuations become yellow'.

On 13 November, Latham again reported on Forbes's progress. He recorded that calomel had restored the patient's evacuations, and that there had been twenty-four hours without any general change in his health. At a moment when he 'felt particularly well', however, 'he felt a tickling sensation at the root of the trachea and up came half an ounce of pure fluid florid blood'. Forbes was immediately given a couple of grains of lead, and Latham was summoned. On arrival at Ventnor, Latham found the patient 'in great alarm, not daring to move or speak, and suppressing the smallest inclination to cough for fear of its producing the haemorrhage'. His pulse measured 90 and was 'firm with some power', his tongue was 'rosy, shining and clean', and there was 'no real cough', and yet there was 'a constant provocation to dislodge something that seems to irritate the trachea low down'. This time Latham's examination revealed a more fundamental problem: 'pulmonary apoplexy in both lungs'. This meant an escape of blood into the air cells and interstitial tissues, classic symptoms of terminal phthisis or tuberculosis. Latham prescribed opium before leaving, making no further reference to the patient's progress.

The Making of 'Heart Latham'

By the 1840s, Latham was well known as a physician, so it is perhaps unsurprising that his expertise was sought in the case of John Forbes. Born in London into a physician family, Latham began his medical studies at St Bartholomew's hospital before being appointed physician to the Middlesex Hospital in 1815. He would return to St Bartholomew's as a physician in 1824. In 1816 he graduated MD and delivered a course of lectures on the practice of physic in London; by 1818 he had been elected a Fellow of the Royal College of Physicians.[6]

Latham's early publications were concerned with epidemic disease.[7] It was not until 1828 that he published his 'Essays on some Diseases of the Heart', the subject with which he would be most closely associated, in the *Medical Gazette*. Latham's interest in heart diseases was encouraged by the recent invention and popularization of the stethoscope by René Laennec, and he regularly made use of the stethoscope in his clinical practice. In 1836 he published his *Lectures on Subjects Connected with Clinical Medicine, Comprising Diseases of the Heart*, in which he dealt with the principles of study and observation, with the methods of auscultation and percussion, and with phthisis.[8]

In 1837 Latham was appointed physician extraordinary to Queen Victoria, a title he retained until his death. He worked as a private physician in London, and resigned from St Bartholomew's because of ill health (he suffered from asthma) in 1841. Four years later he published the work that was to define him as 'Heart Latham' for the remainder of his career, and the book in which he made public the last moments of 'T.A.', or Thomas Arnold: *Lectures on Subjects Connected with Clinical Medicine*.[9] It was this association with cardiac health, and his familiarity with phthisis, that made Latham's skills and experience invaluable for Dr Martin as a consultant for his patients at Ventnor.

The Making of Ventnor

By the 1840s, Ventnor's status as a resort for the sick, particularly for those suffering from tuberculosis or heart problems, was well established. In 1884 James M. Williamson recommended the 'Undercliff' (of which the small town of Ventnor was the centre) for all those 'labouring under affections of the heart, either simple or complicated with pulmonary troubles'.[10] The climate of Ventnor was its main draw for the sick and the recuperating. The virtues of the Isle of Wight's 'Undercliff' had first been extolled by the English physician Sir James Clark in 1830, in a volume dedicated to John Forbes's physician, Dr Martin.[11]

Concepts of 'climate' were crucial to nineteenth-century curative traditions, classified generally in terms of their 'influence or effect upon bodily states, such as bracing, exciting, relaxing and sedative'.[12] There were understood to be four main types of climate, according to Williamson: first, sedative; second, stimulant; third, 'an atmosphere adventitiously impregnated', and, fourth, relative altitude above sea level. Examples of sedative climates—which 'depress[ed] the circulation and weaken[ed] tone'—were Madeira, Torquay, and Jersey. Stimulant climates were believed to be best for pulmonary and cardiac ailments, asserted Williamson, according to his assessment of 693 individual case studies.[13] It was to the 'stimulant' climates that Ventnor belonged, along with St Leonard's and Worthing in Britain, and Cannes, Mentone, Nice, and San Remo on the shores of the Mediterranean.[14]

The medical relevance of Ventnor's climate emerged in the early decades of the nineteenth century. In 1836, inspired by the writings of Clark, his patron, George Anne Martin, along with his surgeon brother, John Baratty Martin, arrived at Ventnor to set up their practice. The fashionable nature of climate-based retreats and the increasing number of people reporting phthisis and heart-disease symptoms meant that they were not short of patients. In 1869 the

Royal National Hospital for Consumption and Diseases of the Chest was established at Ventnor.[15] It was later known as the Royal National Hospital for Diseases of the Chest.[16]

Latham's dealings with Ventnor were apparently not extensive. Between the time of his retirement from St Bartholomew's and his death in 1875, most of his work centred on the small, exclusive medical practice he had established in London. And yet his work in Ventnor placed him at the centre of a number of networks of health and disease connected to the heart and its functions. His influence was seen in the cases of both Thomas Arnold of Rugby and Harriet Martineau, literary individuals linked not only by their residency in the Lake District, but also by a topography of health and heart consciousness that seems to have been a collective and performative phenomenon amongst the creative and cultural elite of the Victorian period.[17]

Examining the medical practice of Peter Mere Latham provides an access to this topography of health and heart consciousness. There are few opportunities for historians to access the dynamics of the therapeutic encounter in the early nineteenth century, and the unpublished casebooks of Latham provide a rare insight. In two volumes written between 1839–40 and 1840–4, we find detailed records of the patients treated by Latham, and the kinds of symptoms for which he was most frequently consulted.[18] This chapter examines the processes behind some of those therapeutic encounters, evaluating what they reveal about cardiac symptoms, experiences, and treatment. It will draw on, and contribute to, the historiographies of subjective experience, of the case study, and of medical therapeutics more generally in the nineteenth century, as well as allowing us to test out some of the assertions about holism made in the previous chapter.[19]

There are, of course, limitations to Latham's casebooks if we wish to access subjective experiences of health and disease. Unlike the letter collections and similar epistolary sources of Edinburgh physician William Cullen, Latham's casebooks provide little in the way of subjective descriptions of patient experience.[20] We do not have the same sense of an interchange between patient and practitioner; nor can we access the languages of health and disease revealed by patients' narratives of experience, however problematic.[21] And yet the structure of casebooks is illuminating when it comes to examining the processes behind the medical construction and interpretation of described symptoms.

What we have in Latham's casebooks is a series of fragmentary encounters, some of which have follow-ups recorded, with the progress of symptoms available to plot and identify. We can determine the priorities made by the physician, and the methods of investigation used to determine the origin and cause of those symptoms. And we are able to access, through the structure of his documentation, the weight given to subjective over objective experiences (or vice versa) and the kinds of diagnostic and prognostic work undertaken. In relation to the heart in particular, we can assess the extent to which cardiac sensations are understood in relation to the entire bodily economy, and how far those sensations (such as palpitations) are indicative of disease, or constitute disease in their own right.

Constructing Case Studies

One of the most striking aspects of Latham's casebooks is their uniformity in terms of case-study structure. Since the publication of pathological anatomy cases in the eighteenth century, the case study had evolved as a specific way of explaining and describing symptoms of disease. It was also one of the prime ways by which knowledge was accumulated, circulated, compared, and legitimated, in order to construct a body of knowledge appropriate for the medical community (as constructed in and through these very processes of intellectual exchange).[22]

According to the categorizing principles of nineteenth-century scientific-knowledge distribution, the case study was an appropriate way to build up evidence, through observed example, in support of the understanding of particular diseases. As Latham put it, 'all our knowledge was originally derived from cases. And cases must still be noted and preserved, and studied, as records of what we know, until we arrive at more general facts or principles than we have yet reached.'[23] This was somewhat removed from the individualistic interpretation of imbalance and interiority allowed under humoralism, but not yet the rigid focus on disease *qua* disease found in twentieth-century interpretations. Latham wrote expansively about the nature of disease and the physician, and he emphasized, despite his focus on the case study, the individually and socially influenced characteristics of afflictions:

Prior to diseases, to their diagnoses, their history, and their treatment prior to them and beyond them, there lies a large field for medical observation. There are things earlier than the beginning, which deserve to be known. The habits, the necessities, the misfortunes, the vices of men in society contain materials for the inquiry, and for the statistical, systematising study of physicians, fuller, far fuller of promise for the good of mankind than pathology itself.[24]

The limitations of pathological enquiry in understanding disease were clear for Latham, for diseases were always socially situated. Moreover, there was a rigidity to the division of the body by pathological anatomization that Latham distrusted: 'To the eye of the anatomist, the vascular system and the nervous system are things apart one from the other. But to the physiologist, the pathologist, and the practical physician, they are always mixed.'[25]

Reflecting (and reinforcing) the drive to know more about particular conditions, but stressing rather more the universality rather than the individuality of disease entities, late-nineteenth-century textbooks similarly defined and categorized the symptoms associated with specific diseases (in much the same way as collected works of pathological anatomy had done). Guidance on cardiac normalities and abnormalities advised practising physicians to think about skin tone, posture, demeanour, and the ways in which heart problems could be detected from the physical deportment and condition of the patient.[26] Within these accounts, there was often explicit reference to this self-conscious regard for categorization, and for the processes by which some form of authoritative medical knowledge could be constructed (usually from the individual to the collective through comparative investigation).

In 1884, for instance, the Edinburgh-based lecturer and pathologist Byrom Bramwell published *Diseases of the Heart and Thoracic Aorta*, a book that was originally delivered as lectures to the author's students in 1883–4. In a chapter entitled 'Clinical Investigation', Bramwell outlined the process by which case studies were taken. Included in that section were the patients' symptomology and some reflections on the examination process.[27] Under the heading 'physical examination', Bramwell advised the collection of 'normal physical signs (i.e. the signs appreciable to the senses—aided and unaided)' and compared these to the 'pathological physical signs' detected upon the patient's body. Before this could take place, however, a detailed 'method of case-taking' had to be established.[28] The 'preliminary facts' about the patient needed to be taken. These included the name, age, sex, and marital status of the patient, as well as his or her occupation, full postal address, and date of admission to hospital. The records that were compiled by Bramwell were clearly more expansive and detailed than those recorded by Latham in his private practice. Nonetheless, the principles of collecting basic biographical facts, to be pored over for clues, and to be compared between patients and over time, were remarkably similar.

The second part of the section on 'physical examination' concerned the 'complaints' of the patient—the subjectively recounted symptoms that had brought the patient to consult the physician. This was to be followed by a history of the present illness: when it started, what were the 'exact character' of the symptoms, and the order of their appearance, and all treatment that had been adopted. In acute cases, students were advised, the temperature of the patient should be taken. After asking about the history of the symptoms, the physician should ask about the 'healthy history' prior to this illness, and the 'habits, mode of life, and general surroundings of the patient'. Finally, a note was made of any relevant family history of illness, especially the 'occurrence of heart affections or of acute rheumatism among near relatives'.[29]

In Latham's account of patient-physician interaction, both objective and subjective assessments were influential. But, 'in matters of feeling', he had cautioned, with explicit reference to the concept of pain, 'we must depend entirely upon what our patient tells us. Every man smarts with his own pain; himself, and nobody else, can say how much.'[30] Some consideration must be made of the patient's subjective observations in all things, though the physician was obliged to persuade him or her of his best judgement about an ailment.

By contrast, Bramwell's account put the physicians' observations foremost. After recording symptoms, they were required to examine the 'physiognomy of the case'. Was there anything 'abnormal' about the physical appearance of the patient, either in 'breathing' or in the appearance of the skin? What was the general state 'of nutrition', or the patient's 'attitude', even his or her facial expression? Physicians were advised to collect information on the 'presence or absence of subjective symptoms referred to' in the heart, or in 'distant organs or parts'.

Finally, and in the case of cardiac patients, the physical examination should note, through percussion, palpitation, and auscultation, the nature and position of the heartbeat, the rhythm of the heart, the character of the heart sounds, its loudness and intensity, and 'its purity'.[31]

After a careful record was taken of the symptoms, physicians were required to make their diagnosis and prognosis, and to prescribe treatment—'hygenic, dietetic, medicinal (general and local)'. In the case of hospital patients, physicians were required to plot the course of the patient's progress, including his or her stay in hospital, the 'mode' and date of the case's 'termination', and 'in fatal cases the record of the post-mortem examination, and an account (where necessary) of the subsequent microscopal examination of the tissues and organs'.[32]

Latham was a hospital physician for many years, and his private casebooks followed a similar principle to that outlined by Bramwell for the Edinburgh Royal Infirmary. However, his patients' biographies were relatively brief. In part, this reflects the fact that they were intended for Latham's use only—in many cases he would have known the personal circumstances of the patient, particularly if he had acted as his or her physician for some time. Where biographical information is given (and it is often just the name of the patient), it is scant. Incidental information might give rise to speculation on the marital status of the men (for women it is clearer, because of the titles given) and occupational categories. Ages were sometimes noted, sometimes not, and the reason for this divergence is not immediately apparent. Despite these differences, each of Latham's case-study entries begins with an introduction to the patient: his or her name, and perhaps an age, was noted prior to the description of symptoms. It is often stated where there is a family predisposition to a particular disease, and the trajectory of the symptoms—when they occurred, how long they persisted, and what treatment had been given with what results—was embedded in the brief discussion of the consultation recorded by Latham.

On 8 July 1838, for instance, Latham was consulted by one Mr Swift, of whom nothing is recorded other than his surname. It is apparent that this was Swift's first consultation with Latham, for a summary of Swift's experiences to date was also recorded. Latham notes that Swift's symptoms had been apparent for four months, and, since the preceding March, when he had given up his 'habitual purgative medicines'. Then it was that Swift 'suffer'd such a dimness of sight that he could neither read nor write'. At that stage he experienced no pain in the head, but some 'numbness' and peculiar sensations in his 'left lower abdomen'. He was also aware of some 'numbness of the left arm'. Over the course of a fortnight, he had subsequently regained the use of his leg and his sight, but he continued to be troubled by itching and weakness in the afflicted regions.

After listening to the patient's personal narrative, Latham would direct the discussion towards his or her overall constitutional health—in the case of men, bowel function, and in the case of women, bowel function and menstruation. Evidently in response to Latham's questioning, Swift reported that his bowels were 'indolent, costive' and 'habitually loaded'. After hearing the patient's story, Latham would begin his physical diagnosis. This normally meant an examination of the patient's pulse, and of the tongue, both of which were crucial indicators of overall health.

The Pulse and the Tongue

Detecting the pulse of the patient through touch was an essential part of the physician's examination. It was also one of the means by which patients could monitor their own subjective experiences and record results in keeping with an increasingly standardized, quantitative means of health assessment. Latham wrote of the pulse as an important indicator of general bodily health, not merely that of the heart. In his 'General Remarks on the Practice of Medicine' (1863) Latham explained that the heart was the most 'sympathetic' of all the organs, and so the pulse 'telegraphs intelligence' of the body 'through every artery that can be felt'.[33] It is interesting that the telegraph had been increasing in use since its invention in the 1840s, and Latham was therefore able to utilize a broadly understood technological metaphor for the distribution and movement of the pulse through the body.[34]

Although there was an established norm from which the qualities of the pulse were seen to deviate, its measurement allowed for individual peculiarities.[35] There was, according to Latham, scope for considerable variation without indicating any structural defect. Yet there were also some specific and quantifiable concerns of which the physician must be aware: these included the pulse's number, frequency, succession, and regularity. Its qualitative measurements were hard, soft, large, and small.[36] In the case of Swift, his pulse was '80 and hard'.

Along with this qualitative and quantitative measurement of the pulse, aural perception of the heart through mediate auscultation (using the stethoscope) was an important aspect of clinical diagnosis. As Latham explained in his *Lectures*, the sounds made by the heart were an important aspect of determining health or disease. In this, the stethoscope was a more reliable indicator than percussion (tapping the chest). In the cadaver, the sounds that emanated from the chest could be like those in the living patient. But the heart's own sounds were produced by its 'own vital movements'. And these could not be taught: 'the sounds, which naturally accompany the movements of the healthy heart, can only be learnt by the practice of listening to them. It is useless to describe them. They are simple perceptions of sense, which no words can make plainer than they are, once the ear has become familiar with them.'[37] Of course, there was an emergent poetics of heart sounds in nineteenth-century culture, which has been discussed elsewhere by Kirstie Blair and others.[38]

Measuring the pulse was not always a reliable indicator to what was happening within the body. On 30 April 1839 one Mr Burgess consulted Latham because he was 'fighting for breath with a cold perspiration pouring from every part of his skin', though his 'pulse was nevertheless strong'. He was prescribed camphor, morphiates, and sulphur; camphor and morphiates were often prescribed in conjunction, to calm and soothe the patient, and to provide pain relief. Sulphur has been associated with shortness of breath, but was a remedy most associated with diseases of the skin.[39] On 3 May Latham visited his patient again, to discover that his shortness of breath had worsened. He 'dozed himself into tolerable tranquillity, but still suffers much dyspnoea'.

Like the pulse, the condition of the tongue indicated the general health and well-being of the patient's constitution. Categories included the colour of the tongue ('pink' being the optimum, 'grey' and 'dull' less desirable). The tongue could be 'moist' or 'dry' or 'clean' or 'loaded' (the ideal being 'clean' and 'moist'). Swift's tongue was not described as healthy, being 'covered with mucous'. The tongues of other patients were described in similar detail, that of Sir W. De Marten, who had just returned from the Continent, being 'whitened and moist with a slight redness at the edges'.[40] There was and is a long

Chinese tradition of understanding the tongue as symptomatic of the body's condition.[41]

Bloodletting, Purging, and Vomiting

For most symptoms, bloodletting, purging, and vomiting were the most common treatments recommended by Latham, along with a range of chemical remedies (such as sulphur) that varied in harshness and severity. De Marten, whose tongue had concerned Latham, had been diagnosed abroad with a liver condition. When he consulted Latham, he complained that he suffered constantly from 'pains of the most excruciating kind' as well as 'obstinate bowels' and 'vomiting'. There was, Latham discovered upon examination, 'some enlargement in the region of the liver'. When he saw his patient again, De Marten complained of 'excruciating pains coming and enduring for hours with tenderness of the scalp'. The habitual sickness meant that he was 'sometimes retching, sometimes vomiting, the vomiting never producing relief except on the condition of a large rejection of green bile'.

Apparently according to the implicit understanding that—as established by ancient humoral medical tradition—there was an internal imbalance within the patient that needed to be rectified, Latham recommended purging for the patient. This only produced 'relief on condition of the quantity evacuated', however, and that quantity was 'very copious and very foul'. On investigating his abdomen, he found that there was hardness and tension on the right, 'gradually passing off into the hypochondrium'. Latham prescribed more purges, though he initially avoided giving leeches, on the grounds that his patient had been prescribed leeches abroad without receiving 'any benefit'. Two weeks later Latham had changed his mind, recording in the case of De Marten: 'leeches used largely'. Apparently in response to the leeches, the bowels discharged 'a large quantity of the usual sort of foul matter and vomited once'. After this evacuation, the patient became 'almost free from pain in the head'. When Latham examined his abdomen, the results appeared even more successful, 'the tension' having entirely 'subsided'.

Most of the patients recorded by Latham had symptoms related to phthisis and to cardiac complaints, the latter of which seemed to increase as the 1840s progressed. This was probably a reflection of Latham's growing reputation as a cardiac physician, of his interests in the field, and of a growing public awareness of the symptoms and possibilities of cardiac disease. There remains little information about the patients, though many seem to have been from the middling and upper levels of society. In those occasional instances where servants or tradespersons were referred to Latham, their treatment was arranged, and presumably paid for, by their masters and mistresses.

Cardiac Cases

When the 27-year-old Fred Maguire consulted Latham in the 1840s, he was concerned that he was experiencing palpitations, characteristic of both 'nervous' and organic disease, though the former diagnosis was normally restricted to women. Although most of his daily activity seemed to pass without trouble, he was concerned that he could not 'make extraordinary exertion in running without inconvenience'. On examination, Latham found that Maguire's heart did, indeed, beat 'with more force than usual', and its reverberations were felt 'beyond the apex but not beyond the praecordial region'. The patient's symptoms were relieved with abstinence from alcohol.

In the case of palpitations and other symptoms—irregular heartbeat, difficulty breathing—it was difficult to distinguish between nervous and organic disease, the categories of which were still in flux in the 1840s.[42] References to a 'nervous' heart nevertheless first came into being in Latham's notes from this time. When 18-year-old Mr Remmet complained to Latham in 1840 of pains 'in the abdomen, varying in situation but chiefly concentrating itself around the left iliac region', the difficulty of disassociating organic and nervous symptoms was clear.

Remmet's physical appearance was, unusually, described by Latham in some detail, suggesting that it was relevant to his diagnosis. There is a reminder of humoral physiology in the careful noting of the patient's 'red hair', 'fair complexion', and 'flushed cheeks'. On examination, Latham found that his pulse was '90, a little sharp', but that his tongue was 'clean'. Remmet's bowels were described as 'capricious', not a phrase commonly used by Latham, and especially unusual when referring to a male patient. After physically palpating the patient's abdomen, Latham found that it lacked 'its usual softness', though there was 'no notable distension'. The patient was unable to 'bear pressure, and is constantly accommodating his posture to the easement of his pain'. These symptoms had lasted for several months, being relieved once, but returning soon after the patient had caught a cold.

Remmet had already seen several physicians about his disorder, several of whom, Latham noted, 'not being able to detect his disease, have concluded it to be nervous'. Latham disagreed with this diagnosis. Like his peer, the eminent physician Thomas Watson—whom Remmet had also consulted, and who appeared to be in regular contact with Latham over individual cases of cardiac dysfunction—Latham believed that there was indication of some 'chronic disease, probably tuberculosis'.[43] After consultation with Watson, Latham agreed to watch the patient closely over the next few months, to await 'further indications'.

There were several instances when cardiac symptoms were acknowledged to be linked to emotional distress. On 1 March 1841, Mr Henry Coulthurst complained of pains in his chest and palpitation 'especially at night [and] with a sense of fainting'. These symptoms had continued for many months but were 'aggravated lately by the death of his brother—a hardworking, sedentary lawyer'. Coulthurst's bowels were regular, his urine 'loaded'. Also relevant were the patient's habits and mode of existence: he frequently dined late and ate heartily. His drinks included 'soup & wine & porter & afterwards tea—& finally perhaps gruel'. Latham instructed him to omit his porter and to take tea but without the gruel. Three days later, Coulhurst's health seemed to have improved: there was a very slight murmur 'accompanying the systole at the basis' of his chest, but 'not one of his discomforts remain'. His urine had become 'clear'. Impressed by his patient's progress but still concerned that he might be working too many hours, Latham instructed him to work less, and to take at least two days' 'relaxation' from business per week.

The discussion of 'habits' is important, because it was well understood in nineteenth-century medical practice, as it had been since the classical period in the realm of medical theory, that 'general lifestyle' affected the condition of the heart, as well as the other organs, and the overall health of the patient: the kinds of food and drink that were taken, loss of sleep, disturbed sleep, and a range of 'troublesome wants, such as frequent micturation'.[44]

In common with classical doctrines on health and disease, therefore, Latham's examinations revealed that insufficient work was as
detrimental to the health as excessive work. On 10 April 1841,
Mr Edwards, a 58-year-old gentleman whom Latham had treated
seven years previously when suffering from 'insanity', consulted Latham. He said that he had been well until the previous few days, and
there was no discussion of the symptoms that had led him to consult
Latham, although he did 'speak of extraordinary dreams which he
will not reveal'. On examination, Latham found his head and skin
'hot' to the touch, and his skin 'generally hot' all over his body. His
tongue was 'furred', his pulse 'a little hard', and his abdomen 'full'.
Edwards's psychological state was clearly of concern to Latham, as
Edwards reported 'of the necessity of his going into Yorkshire to get
his father's will altered', despite the fact that his father had died 'long
ago'. His diagnosis seemed to focus on the 'habits' of Mr Edwards.
Although he 'slept well', he did not engage his mind appropriately
for optimum health: 'his habits were formerly those of a man actively
engaged in business: at present they [sic] dawdling and idle having no
employment.'

   Most patients referred to Latham suffered from phthisis and, by the
later 1840s, cardiac symptoms. But in each case the heart's function
was seen to be responsive to the general and overall health of the
patient. Many, like Miss Raymond Baker, who consulted Latham
in April 1839, generally 'relie[eved] of [their] symptoms following
relief of the bowels'. There is a growing body of literature on the
bowels and the abdomen in the eighteenth and nineteenth centuries,
and there is ample ancient precedent to indicate that the action of
the bowels (as related to incretion and excretion, and as one of
the 'non-naturals'), has long been seen as relevant to an individual's
emotional and physical health.[45] In understandings of cardiac function,
the maintenance and operation of the patients' bowels seemed crucial
to Latham's assessment.

The Heart of Holism

As the above examples have shown, Latham's practice maintained
a holistic, often Galenic, attitude to illness and disease. Patients
continued to be bled, blistered, and purged, whatever shifts were

underway in 'scientific' understandings of disease more generally.[46]
The body's excretions—from sweating to stools, from mucus to
blood—could demonstrate physical problems deep within the body
that could be related to heart complaints, or produced by cardiac
problems. This is at odds with an understanding of the nineteenth
century as a transitional period, when patients' narratives of disease,
and holistic interpretations of the mind–body relationship, were
outmoded in favour of newly emerging 'scientific' models.[47]

Latham's attitude towards the holistic framework of medical practice, and towards disease diagnosis, is explicit in much of his writing.[48]
In a section entitled 'practice', Latham writes of the development of
clinical medicine during his lifetime, and of its tendency to specialize
and to localize disease. Despite some positive or 'useful results', he
complained, this process had 'a tendency to narrow our views, and to
cripple our practice by setting up as many several pathologies within
the body as there are several organs':

> Yet no sooner do the diseases of separate parts come to be treated, than they
> begin to claim their place in a common pathology. We cannot reach them,
> and apply our remedies directly to them, in the isolated spots wherein we
> find them; but if they are to be reached, and treated at all, it must be through
> the vascular system, or through the nervous system, or through the digestive
> and assimilative system. For these are the common agents of life and increase,
> both healthy and unhealthy, and the common channels both of food and of
> medicine.[49]

The links between Latham's attitude towards medical practice in the
1840s, and the 'non-naturals' philosophy of Galen and the ancient
world, is apparent. And yet Latham has been virtually ignored in
nineteenth-century accounts of medical practice, particularly in the
realm of therapeutics. Although cardiologists have included sections
on 'Heart Latham', generally in relation to his practitioner role as part
of the 'new science', his work as a physician has been neglected. As
this brief examination of his casebooks has shown, however, 'Heart
Latham', the oft-cited forefather of modern, scientific, cardiological practice, showed the same adherence to holistic precedent and
tradition that medical historians have identified elsewhere.[50]

By the 1840s, then, medical practitioners had yet to gain any
monopoly over the experience and understanding of cardiac function
or dysfunction. Nor did scientific interpretations influence the languages of emotion and the heart that were increasingly embedded in

literary and artistic understandings of truth, integrity, and intelligence. As we will see in the following chapter, and in the writings and biographical experiences of Harriet Martineau, the mid-nineteenth century saw heart disease take on a new dimension, at least among the literary and cultured elite.

Rather than being evidence of mechanical dysfunction, heart disease was arguably redefined around mid-century, and in certain social circles, as evidence of a heightened emotional sensibility, continuing the link between emotions and the passions we have traced back to the ancient world.[51] It is no coincidence that one of Martineau's physicians (and the individual who, along with Sir Thomas Watson, would debate the existence of Martineau's heart disease) was Peter Mere Latham, his dominance as a cardiac specialist having increased by the 1850s. As seen in the following chapter, Latham's negotiations with Martineau over the nature of her cardiac symptoms, seldom analysed as part of any existing historiography, tell us much about the shifting meanings of heart disease in the midst of Romanticism. It also provides a counterbalance to the scientific materialism discussed above.

6

The Heart of Harriet Martineau

> I had been kept awake for some little time at night by odd sensations
> at the heart, followed by hurried and difficult breathing... The
> disturbance on lying down increased, night by night. There was a
> *creaking* sensation at the heart (the beating of which was no longer
> to be felt externally); and, after the creak, there was an intermission,
> and then a throb. When this had gone on a few minutes, breathing
> became perturbed and difficult; and I lay till two, three, or four
> o'clock, struggling for breath. When this process began to spread
> back into the evening, and then forward into the morning, I was
> convinced that there was something seriously wrong.
>
> *Harriet Martineau's Autobiography* (1855)[1]

In earlier chapters, this book has considered the shifting status of the
heart from the classical period to the nineteenth century, noting its
increased medicalization as an aspect of the rise of scientific, rational
medicine, and the emergence of mechanistic ways of understanding
the mind–body relationship. Of course, this is not the only story.
Hegemonic beliefs or structures are not produced in a unidirectional
manner; on the contrary, counter-models and counter-memories are
an important aspect of the process of historical change.[2] The most
important counter-model to the rationalization of the heart in the
nineteenth century has yet to be explored: how it was that, at the
very same time that the heart was becoming secularized in medical
theorizing as a materialistic entity, subject to decay and divested of its
previously spiritual significance, it was also becoming *more* sanctified
than ever before.

The rise of Romanticism, with its emphasis on the individual,
on authenticity and truth, and on the prioritizing of emotion over

Harriet Martineau.

Fig. 6. Harriet Martineau. Wood engraving.
Source: Wellcome Images.

reason (largely as an antidote to the preceding and rational Augustan
period), provided an alternative to the medicalized heart. The heart of
Romanticism was a heart filled with feeling, to be shared with like-
minded people or clutched to one's breast as evidence of one's own
individuality. Like the sacred heart of Catholicism that had flourished

in post-Reformation Europe, the Romantic heart spoke essential truths about some universal human 'nature' as well as reflecting traditional beliefs about the relationship between emotions and the divine.[3] For writers, as for theologians, the heart possessed a unique source of wisdom, intellect, and feeling. In this context, 'heartfelt' sentiment revealed much about the identity of the person experiencing it: amongst the middle classes in particular, extremes of emotion began to indicate a sensitivity that betrayed one's elite and emotionally sophisticated status. And yet the history of the embodied heart is a neglected aspect of the history of sensibility.[4]

This chapter explores these issues more fully through an examination of the case of the writer, philosopher, and political activist Harriet Martineau, who promoted the image of the heart as emotional centre, both in her work and in her personal life (Fig. 6). It builds on recent work into the centrality of the heart in Victorian fiction by highlighting its importance to concepts of subjectivity and self-fashioning—promoting some form of embodied 'identity' or 'experience' that, though problematic, has become central to modern literary and medical analyses.[5] In this context, Martineau's body provides evidence of the widespread relevance of the pathological heart in nineteenth-century medical debates, as well as of the preservation of emotion in literary formulations of the heart.

Martineau and Medical History

In the summer of 1839, Harriet Martineau fell ill during a trip to Venice. Martineau had suffered from a number of physical ailments since childhood, and was profoundly deaf. Yet her collapse in Venice aged 37 marked a new state of decline. Although she was to live until she was 74, the rest of her life was to be spent in poor health, a general debility of body punctuated by periods of extreme pain and suffering. Martineau's extensive symptoms were ultimately attributed by the medical profession to a uterine disorder—more precisely to a large ovarian cyst.[6]

Martineau's frequent speculations on the cause and nature of her suffering have been explored by historians of Victorian gender, literature, and medicine.[7] We know that she rejected many orthodox medical interpretations, and that she was humiliated by, and angry at, the publicization of her symptoms by Dr Greenhow, Martineau's physician and brother-in-law. Although it was Greenhow who first suggested, and arranged for, Martineau's sessions with a mesmerist, he was responding to, and seeking to dispute, her subsequent claims about the experience—most explicitly, her claim that she had been cured by the increasingly controversial practice of mesmerism.[8] Greenhow's pamphlet was graphically detailed.[9] It included Martineau's menstrual history and discharges from and irregularities in her vagina, all of which (and 'not even written in Latin—but open to all the world!') mortified Martineau, and caused subsequent conflict within the extended family network.[10]

From the 1840s, Martineau became more aware of the politics of health. In 1844, while confined to her sickbed in Tynemouth, she wrote *Life in the Sick Room*, a series of essays, philosophical and psychological, on how to be an invalid. And yet, between 1845 and 1855 Martineau appeared to make a good recovery. She settled in Ambleside, in the Lake District, where she continued to write prolifically about women, slavery, and the political economy. She travelled to the Middle East, and in Scotland and Ireland, as well as riding, walking, hiking, and climbing in the Lakes. She also maintained her literary and political networks by correspondence with other influential writers and philosophers, including Matthew Arnold and William Wordsworth. Yet the onset of a further period of ill health in 1855, this time in the form of a cardiac complaint, was both sudden and dramatic.

It was at Ambleside, after experiencing the cardiac symptoms noted in the epigraph above, that Martineau wrote her *Autobiography*, claiming that she was in anticipation of imminent death.[11] In this work, and in her simultaneous correspondence with friends, Martineau described her physical decline as a consequence of serious heart disease. She became 'convinced' that there was 'something seriously wrong' with her heart, a conviction that was apparently proved correct when she received a terminal diagnosis from Peter Mere Latham and Sir Thomas Watson, two of the country's most eminent cardiac specialists. When she was aged just 53, Martineau's heart had become simply 'too feeble for its work'.[12]

Although she lived until 1876, Martineau recorded her subsequent death in a self-penned obituary that was tagged to the *Daily News'* introduction to Martineau's *Autobiography* in 1876: 'Her disease was deterioration and enlargement of the heart, the fatal character of which was discovered in January, 1855. She declined throughout that and subsequent years, and died.'[13] It is clear that, in preparing herself, and her readers, for her death, Martineau viewed heart disease as the logical end point of her life. And yet the heart of Harriet Martineau is a neglected aspect of medical and literary historiography.

Apart from important feminist acknowledgement of her political and social contributions to the nineteenth century, through her writings on the abolition of slavery and the economy, literary medical historians have focused mainly on the narrative construction of Martineau's gynaecological illness, using the rich evidence provided in her *Autobiography* and *Life in the Sick Room*. Susan F. Bohrer, Maria Frawley, and others have used these sources, along with Martineau's published correspondences, to examine the strategies of self-representation open to women in Victorian culture.[14] These works have shed new light on concepts of maternity, disability, femininity, and the nature of the 'public' in nineteenth-century Britain. A less-explored aspect of historical investigation uses Martineau's testimonies to consider the meanings of medical knowledge, as seen in both Alison Winter's and Roger Cooter's accounts of Martineau's conflicts with the medical profession over mesmerism.[15]

Despite such a rich historiography, and with few exceptions, Martineau's heart disease has received little attention. Some biographers have overlooked discussions of this diagnosis, whilst others have rejected it out of hand, or remarked on it simply as one in a series of strange illnesses to which Martineau—portrayed as a contradictory and difficult individual—succumbed. In the latter instance, accounts tend to be patriarchal and patronizing. In *Harriet Martineau: A Radical Victorian*, for instance, R. K. Webb refers in passing to his subject's fear of heart disease, 'which she did not have', in a typically dismissive fashion: 'like so many Victorians, she revelled in illnesses which were painful and distracting, though usually not quite so bad as she made them out.'[16] More sophisticated analyses have examined Martineau's cardiac complaint as part of a complex debate with physicians over the nature of medical authority and the value of subjective experience.[17] Yet there has been no systematic analysis of Martineau's status as a heart-disease patient, or of the reasons why this might have provided Martineau with a valid vehicle for self-fashioning.

By focusing on the heart of Martineau, this chapter seeks to provide such an analysis. It picks up several existing strands in historiographical research, including Martineau's narrative construction of her illness, and the relationship between medical authority and the 'patient'. It offers a speculative interpretation of the heart of Harriet Martineau in regard to her relationship with her physicians, and in the context of Victorian attitudes towards the culture of the heart best identified by literary theorists.[18] It also uses several previously unpublished correspondences, including those between Martineau and the cardiac physician Peter Mere Latham. Latham was physician extraordinary to the Queen from 1837. His cardiac expertise was established by his *Lectures on Subjects Connected with Clinical Medicine, Comprising Diseases of the Heart*.[19] Martineau's correspondence with Latham is looked at here in conjunction with her published letters, her *Autobiography*, and contemporary medical reports and treatises.

Situating Martineau's case in the context of contemporary medical and literary attitudes towards the heart, emotions, and disease, this chapter argues that Martineau's self-declared identity as a heart-disease patient was performed by a sleight of hand whereby she set up her own experience in opposition to, yet also appropriated, medical authority. This is a different matter from disagreeing with her physicians' diagnosis on the basis of her own 'lived experience', a characteristic that historians have often associated with Martineau's illness.[20] Martineau did not claim to know something *more* than her physicians; she harnessed their authority in order to declare herself a heart-disease patient. Her physicians were also the most respected authorities available.

Along with Latham, Watson was nationally recognized as a key cardiac specialist. In 1859 Watson was also appointed physician extraordinary to the Queen, and he served as President of the Royal College of Physicians from 1862. His *Lectures on the Principles and Practice of Physic* was published between 1840 and 1842, and it was the chief English textbook of medicine for the next thirty years.[21] In medical and literary circles then, the opinions of Latham and Watson would have been significant and influential, validating her subjective experiences. While she was quite prepared to downplay the gynaecological diagnoses of Greenhow and others (at least in public), Martineau claimed that the diagnoses of Latham and Watson accorded with her own: all agreed that her heart was 'too feeble for its work'.

Yet an examination of the correspondence and medical debates surrounding her case suggests that this was far from accurate. At the same time that Martineau was writing to her friends to tell them that Latham and Watson had diagnosed her with a fatal heart complaint, those same physicians were assuring her that her symptoms were *not* serious, and were related *not* to organic heart disease, but to a related and general debility of the body caused by an ovarian cyst. This appears to have been a deliberate act of rewriting; Martineau rejected the public persona of an ovarian-cyst sufferer, tainted as it was with public humiliation as well as women's reduction to biology, and took instead the mantle of heart-disease patient. This would provide Martineau with an experientially explicable, and more publicly palatable, window onto her suffering. For not only was heart disease a less politically and personally distasteful condition than an ovarian cyst; it also tapped into a rich vein of medical and literary writings that attributed to the sufferer emotional sensitivity, intellect, and poetic aptitude.

The Heart of Martineau

Martineau was 52 years old when she began to experience 'odd sensations' in her chest. These sensations were followed by periods of difficult breathing and even—when reading in the daytime—some apparently connected difficulties with her vision.[22] When Martineau began to experience this 'hurried and difficult breathing', and a '*creaking* sensation of the heart', she resolved to contact Latham.[23] This decision was made, Martineau noted in her *Autobiography*, with the full support of her family. She reported that she was fully prepared, even eager, to take on whatever diagnosis might emerge, recording how 'that honest and excellent physician knew beforehand that I desired . . . to know the exact truth, and he fulfilled my wish'.[24] As was usual for the period, Martineau initially became involved with Latham through a correspondence relationship, and many of Latham's letters to Martineau survive. I can find no trace of Martineau's own letters to Latham. These may have been casualties of another aspect of Martineau's self-fashioning; in 1843 she ordered all her correspondents to destroy her letters to them, upon pain that she would never write to them again.[25]

In his first letter to Martineau, dated 12 January 1855, Latham was evidently responding to the symptoms outlined in the epigraph. He advised that without examining Martineau he could not 'see through her case so clearly' as he would wish, though he was happy to offer some 'guide to treatment'.[26] On the basis of her letter, however, he could offer her some reassurance; Martineau's symptoms were *not* consistent with those associated with heart disease:

Whatever 'the creak, the stop and the thump' may mean [he wrote], they can hardly *in you* mean organic disease of the heart. To walk 7 or 8 miles without inconvenience; to drink port wine with *very good* effect, and to obtain 'a most comfortable day' from 12 drops of Battley's Laudanum[27] are enough to almost abolish any evil suspicion I might have from symptoms immediately referable to the heart itself.[28]

To ease Martineau's symptoms, and to learn more of her condition by seeing how it responded to treatment—'the effect of the remedy often serv[ing] to interpret the disease'—Latham advised that she take a 'very mild opiate (1/8 of a grain), in combination with ammonia'. This she was to take with water every six hours for the space of a week. During this time, Latham cautioned, Martineau was to ensure that she was not constipated, that she took moderate exercise and drank a little wine.

On 18 January, less than a week later, Latham wrote again to Martineau. Her symptoms had apparently worsened since his previous correspondence. Latham was reluctant to 'strike any hard blows in the dark', so instructed her to 'give up altogether' the treatment that he had previously recommended: 'From what your letter of today tells me of your present condition it will not be safe for me to venture further upon your treatment without seeing you'. Yet Latham instructed Martineau not to visit him, advising against travel to London 'in this cruel weather'.[29]

Soon after her receipt of this letter, Martineau arrived in London, staying at the lodgings of John Chapman, physician and editor of the *Westminster Review*.[30] Her rationale for this choice of lodgings was explicit: she 'felt it so probable that I might die in the night' that she refused to go to the house of her 'nearest friends, or of any aged or delicate hostess'. At the Chapman's residence, 'all possible care would be taken of me without risk to anyone'.[31] It was at these lodgings that Latham had first visited Martineau on the day after she arrived in London.

Martineau later described that examination to her friend Maria Weston Chapman. Martineau had met Chapman, a prolific American abolitionist, during a visit to Boston in 1835. Martineau requested Chapman to conclude the final part of Martineau's *Autobiography* after her death. Chapman agreed, and the result was the *Memorials*, appended to the third volume of the *Autobiography*. This contribution was explicitly influenced by the letter she received from Martineau on 24 January 1855. In it, Martineau revealed that 'the first man for heart complaints' had just 'made a long examination [of Martineau's chest] by auscultation'. He 'did not attempt to conceal the nature and extent of the mischief'.

From his being unable to *feel* the pulsation of the heart in any direction, while it is audible over a large surface, he believes that the organ is extremely feeble in structure—'too weak for its work'—and very greatly enlarged.[32]

With expressed regret, therefore, Martineau informed her friend she was 'mortally ill', having suffered some months from 'what now turns out to be organic disease of the heart'. The disease being increased by 'the anxiety and fatigue of the autumn', there was no knowing how much longer Martineau had to live.

According to Martineau's *Autobiography*, and to at least one of Martineau's biographers, Latham urged her to consult another physician who was then acting as his *locum*.[33] One week later, on 31 January 1855, Martineau did pay a visit to the physician, Thomas Watson. Martineau reported the outcome of this meeting in her *Autobiography*. She recalled how Watson's opinion,

formed on examination, without prior information from Dr Latham or from me, was the same as Dr Latham's. Indeed the case seems to be as plain as can well be. It appears that the substance of the heart is deteriorated, so that 'it is too feeble for its work'; there is more or less dilatation, and the organ is very much enlarged.[34]

Even before Martineau had left London, she found herself subject to 'the sinking-fits which are characteristic of the disease'. It was 'perfectly understood by us all that the alternative lies between death at any hour in one of these sinking fits, or by dropsy, if I live for the disease to run its course'.[35] Although she had been anticipating this prognosis before seeing Latham or Watson, Martineau found herself 'rather surprised that it caused so little emotion in me'. She went out immediately to visit a friend, 'to tell her the result of Dr Latham's visit; and I also told a cousin who had been my friend since our school days'. While dressing for dinner, Martineau recalled, she experienced 'a momentary thrill of something like painful emotion...not at all because I was going to die, but at the thought that I should never feel health again'.[36] Martineau subsequently informed her family of her impending demise, and rewrote her will.[37]

From this time on, Martineau lived as a woman dying from organic heart disease. She took on that mantle with little trepidation, describing herself—after years of ill health—as 'more than ready...even joyful in the prospect of sudden departure'.[38] She put all 'her affairs' in order 'as soon as Dr Latham's warning was given', and was quite prepared to die. In her contribution to Martineau's *Autobiography*, Chapman recalled how, a year after this damning diagnosis, Martineau was still regularly, and publicly, 'subjected to very severe suffering':

The frequent recurring of suspense of the heart's action was very alarming. Her recovery from each attack seemed at the time as doubtful as resuscitation after drowning. 'Really and truly', said her friend Lord Houghton, who was accidentally present at one of these sudden seizures, 'one may use St Paul's words, "she dies daily"'.[39]

Martineau and the Medical Profession

Martineau's reporting of her heart disease seemed uncontroversial. And yet it does not quite sit with the available evidence, much of which is, by virtue of the piecemeal nature of references to Martineau's heart disease, suggestive, rather than conclusive. If we turn to the letters written by Latham between 1855 and 1857, for instance, the period when the physician was apparently treating Martineau for her fatal cardiac complaint, we find no mention made of any diagnosis of heart disease. Instead, Latham repeatedly recommends the use of opium, to relieve Martineau's discomfort. The remainder of his advice concerned her experience of persistent 'neuralgic' pains (which Latham believed to be a side effect of the opium, or else symptomatic of actual disease elsewhere in the body).[40] This latter comment suggests that Latham believed Martineau's cardiac symptoms and difficulty in breathing to be a functional condition, derived from and influenced by her gynaecological complaint.[41]

This possibility is supported by Latham's single reference to Martineau's heart in a letter to her niece, dated 25 May 1855. In response to a letter from Maria Martineau, Latham suggested that Harriet 'make trial of certain steel drops with the view of sustaining that weak heart, which is so ready to flag and falter'.[42] Fifteen steel drops were to be taken in water three times over a twenty-four-hour period. Should this treatment 'disagree' with the patient, by 'offending' the stomach, for instance, the remedy should not be continued. Latham's acknowledgement, here, of Martineau's functional heart symptoms did not mean that the physician believed her to suffer from heart disease. Latham apparently believed that Martineau's 'weak' heart was a product of her age and her overall health, and, more specifically, that it originated from a related dysfunction.

In nineteenth-century medical theory, the transmission of symptoms from one part of the body to another was perfectly commonplace. In part this was a nervous characteristic, quite literally—the transmission of symptoms from one site in the body to another through the brain and central nervous system.[43] Latham's earlier mention of 'actual disease elsewhere in the body' is therefore important. For in 1855 Latham wrote again to Martineau. This time he was responding to Martineau's concerns about her increasingly large and swollen abdomen, which she feared was not caused—as had been claimed—by dropsy.[44] He advised that she was 'right' to be concerned, and that he did not believe Martineau's symptoms to be 'dropsical'.

When Latham had examined her abdomen some months earlier in London, he reminded Martineau, he had found it 'enlarged and hard', symptomatic, he believed, of an extensive 'tumour'. When a tumour grew to fill the abdomen, Latham explained, it often appeared to disappear because it was no longer detectable. Yet 'so far from disappearing it has been increasing all the while—and as it has increased it has lost its distinctness'.[45] The belief that Martineau was suffering from a tumour situated in the abdomen was repeatedly suggested by physicians around Martineau, including Thomas Watson.

Martineau consulted Watson at the advice of Latham on 31 January 1855, when Watson was working as Latham's *locum*. As described above, Martineau reported in her *Autobiography* that Watson had concurred with Latham's fatal diagnosis, telling Martineau that she suffered from fatal heart disease. And yet, again, Watson's own record tells a different story. Watson's account needs to be understood in the context of a long-running dispute between Martineau and physicians like Greenhow over mesmerism, on the one hand, and, on the other, Martineau's more general dismissal of the abilities of the medical profession.

After the publication of Martineau's *Autobiography*, Watson wrote to the *British Medical Journal* to complain of the misrepresentation of Martineau's case. With reference to extensive notes taken at the time of the consultation, Watson recalled how Martineau had complained to him of 'intermissions of the beats and subsequent boundings of the heart, felt by her very disagreeably, with flutterings and bumps'.[46] Under examination by auscultation, Watson found the 'pulsations of the heart noisy, and audible over a large portion of the chest; but there were no murmurs attending its action, nor any other evidence of organic disease'.[47]

While Martineau had claimed that Watson declared her to be suffering from an enlarged and dilated heart, therefore—one that was 'too feeble for its work'—Watson denied this claim. He reportedly told Martineau that her heart was in the condition that one would expect in a 52-year-old woman. At the age of 63, he continued, he had 'sufficient experience, in my own person, of these disagreeable flutterings and intermissions of the heart and pulse, lasting sometimes for days together'.[48] Had he had any misgivings about the condition of her heart, had he believed that there were 'flaws in the mechanism of the heart' that needed 'careful management', then he would have advised her, or a member of her family, as a matter of urgency. Yet 'in Mrs Martineau's case there was no such obvious rift, and I, therefore, affirmed to her that her life was in no immediate danger'. He believed, moreover, that she would have received 'a similar opinion from Dr Latham, than whom no physician at that date was more competent to form a correct judgement about affections of the heart'.[49] Watson believed that Martineau was unwilling to receive his diagnosis, however, as the patient 'plainly distrusted, or rather she disbelieved, my reassurances, looking upon them, I fancy, as well-meant and amiable attempts to soothe and tranquilize a doomed patient'.[50]

If Watson did not diagnose Martineau with heart disease, what was his opinion as to her condition? The notes made at the time of consultation were subsequently made available to the British Medical Association. They describe in detail Watson's examination of

Martineau's chest and abdomen, and his conclusion that she was suffering from 'a large pear shaped indolent tumour, reaching as high as the lower part of the epigastrium'.[51] Greenhow concurred, perhaps seeking to defend his medical reputation by reminding readers that he had also identified Martineau as suffering from a gynaecological complaint. He reported that he knew she had been declared free from heart disease by both Latham and Watson, but that 'she nevertheless maintained and asserted her conviction that she would soon die from that cause'.[52]

In Martineau's discussions with her physicians, and in the testimony found in the *British Medical Journal*, it is clear that she struggled with medical authorities over the interpretation of her own cardiac symptoms. On the basis of post-mortem evidence, Martineau's ovarian cyst was seen to have forced her stomach into the thoracic cavity, arching the diaphragm and impeding the action of the heart and lungs.[53] So we have a medically sanctioned and mechanistic explanation for the physical symptoms experienced by Martineau. As with any retrospective diagnosis, however, this fails to address the meanings that Martineau placed on her own experiences, and the politics of the heart and its diseases in Victorian culture.

Why did Martineau choose to view her extensive symptoms as a product of heart disease, rather than as related to her gynaecological condition? After all, as Cooter has observed, Martineau did privately acknowledge her 'tumor' to be the source of her ill health in a letter to John Chapman written after February 1855—in other words, *after* the time that she claimed Latham and Watson had made their terminal diagnoses.[54] As Anka Ryall suggests, this 'cyst', removed, preserved at autopsy, and later circulated amongst the medical profession, 'represented a secret that Martineau, while alive, had carefully guarded'.[55]

In a detailed analysis of Martineau's relationship with mesmerism, Cooter has suggested that downplaying her cyst and highlighting her alleged heart disease would have preserved Martineau's earlier demonstrations of the efficacy of mesmerism; it cured her gynaecological symptoms. This may be correct, but I suspect there were further, complicated factors at work. First, it was against middle-class and gendered sensibilities to discuss gynaecological conditions: we know that Martineau described herself as humiliated by Greenhow's public report. On a political level too, to admit publicly to a tumour would have meant that Martineau succumbed to a 'female malady' at a time when women were increasingly reduced to their biology, and gynaecological specialism was proving one more means by which the male medical profession dominated and excluded female (and lay) knowledge.[56] This problem is acknowledged by Cooter, who notes Sir Charles Mansfield Clarke's statement that for a physician to treat women with any reference to her own perception of her illness was like expecting 'to remove ascarides [round worms] from the anus by making application to the nostrils'.[57]

To Martineau's subsequent regret, Mansfield Clarke had examined Martineau when she was in Tynemouth, and supported Greenhow's diagnosis. Clarke's comments make apparent the reasons why Martineau may not have been enamoured with the medical profession. It should also be clear, from the above discussion, why succumbing to an ovarian tumour might have been publicly unpalatable for Martineau. But why was heart disease so much more acceptable, even idealized? To understand this we need to consider the role of the heart and its diseases in nineteenth-century culture. Some detailed examination of one of Martineau's own fictional writings, moreover, deserves consideration.

Victorian Culture and the Heart of Emotions

At the time of Martineau's writing, the heart held an iconic status as an organ of emotional integrity, truth, and honesty. Many of these cultural meanings were passed down from classical precedent.[58] Yet, as Kirstie Blair has demonstrated, there was a renewed shift towards the heart in Victorian literary culture, in terms of both heart-centred imagery, and its links with emotion and authenticity (the values later associated with the Romantic movement).[59] In nineteenth-century fictional writings, therefore, including those by Martineau herself, the heart functioned as a symbol of intense sensibility and feelings.

Along with the writings of Elizabeth Barrett Browning (see, for instance, Verse V of *Sonnets from the Portuguese*) and Christina Rossetti (whose 'A Birthday' reported that 'My heart is like a singing bird'), a useful example of this phenomenon is found in Martineau's first novel, *Deerbrook*, published in 1839.[60] Its subject is the fortunes of the Ibbotson sisters, Hester and Margaret, who arrive at the village of

Deerbrook to stay with their cousin, Mr Grey, and his wife. Margaret attracts the attention of the local apothecary, Edward Hope, though he is persuaded to marry the beautiful Hester, largely because she believes that he is in love with her. It is a miserable marriage, overshadowed by Hester's jealousy and a series of misunderstandings between Margaret and her own lover. The physician's fortunes take a turn for the worse when he is accused of grave-robbing, and he succumbs to a near-fatal fever. Eventually, and according to convention, health and order are rehabilitated, along with the marriage of Hester and Edward.

Discussed by many critics as one of the first 'domestic novels' of the Victorian era, Deerbrook provided a sentimentalized account of the meanings of love and affection between siblings, acquaintances, and lovers. Central to the symbolism of its narratives of love, authenticity, and choice, hearts had vitality and morality of their own in Deerbrook. They 'danced' and 'sank' under extreme emotions. Individuals became 'heavy'-hearted when grief-stricken; had 'cheerful' hearts, when optimistic. Women, in particular, possessed 'kind' hearts, and thus concern for others, while it was conversely possible to be 'hard'-hearted, and immune to their plight.[61] Over time, hearts became softer or harder, depending on experience. As expressed in the final chapter of the novel, the male protagonist 'had really gone through a great deal of anxiety and suffering lately, and his heart was very soft and tender just now'.[62]

Hearts were also connected to the mind in Deerbrook, and to reason through a symbiotic rather than a combative dynamic. The possession of a 'heart and a conscience' was important to humanity, and 'sympathy within [one's] heart and mind' ideal.[63] Yet hearts also possessed knowledge that was unmediated by human error. To know oneself, or one's subject, was to 'learn by heart'; to follow one's truth was to 'follow' one's heart. In figuring as emotional receptacles, hearts were filled or emptied by degrees of feeling, and characters became 'heart full' or 'single hearted' according to the object of their affections. Hearts also changed shape, depending on the emotion that they expressed. As in seventeenth-century discourses on emotion physiology, hearts 'swelled' when they had secrets to impart; and 'inflated' with pride.[64]

The truthfulness of the heart was physically retained in its structure, for it embodied memories and experiences; it remembered words and tones of speech.[65] Questions therefore 'struggled' in the heart, and the heart physically resisted the suppression or avoidance of its truth.[66] To live untruthfully, and to be distressed 'at heart', was to invite unhappiness or even disease. Little wonder, then, that the actions and sensations of the heart were subject to a series of specific physiological effects. Characters were reported as 'light hearted' when not serious, or 'bitter' and 'heavy' at heart when disappointed. Hearts became 'heavy' and 'dismayed', 'sick' or 'affected' by experience, so that they 'leapt up' (in joy) or 'trembled' and 'beat' in fear and anticipation.[67] When extremely distressed, hearts 'throbbed' painfully, or were 'weighed down by grief'.[68] Any extreme weight afflicting the heart (as in cases of depression and disappointment) constricted one's breathing and caused breathlessness, even palpitation, symptoms also associated with heart disease.[69]

Beyond its physical structure, and at another level of analysis, hearts signified intimacy; metaphorically to share the contents of one's heart was an affirmative action, as opposed to being isolated or lonely (when the heart felt 'wringed' in grief or was 'breaking with its loneliness').[70] The heart needed nurture to stay well; without kindness and affection (indeed, without intimacy), it weakened. How ironic, then, as one character exclaimed in *Deerbrook*, that 'we cannot see into one another's hearts'. For on some level we must always be alone: 'what lies deepest in [our] heart' is ultimately impenetrable by others.[71]

A Fashionable Disease?

This physical siting of emotions in the heart was characteristic of broader Victorian poetic invocations to the heart as the centre and symbol of creativity and truth.[72] In nineteenth-century literary culture, as in earlier periods, the heart functioned as the centre of emotional experience, filled or emptied with hope, joy, fear, anger, and love at the psychological flick of a switch. As Blair has identified in her analysis of Victorian poetics, however, there was a renewed shift towards the heart in nineteenth-century literary culture, in terms of both heart-centred imagery, and its links with emotion and authenticity (the values later associated with the Romantic movement).[73] This shift emphasized not only the creative symbolism of the heart, but also its pathological potency.[74] In medical and literary terms, the heart was affected, and effected, by extremes of emotions. Excessive emotions could cause disease in the whole body, including the heart; heart disease could cause extreme emotional disturbances.[75]

Heart disease was a much-debated topic amongst the medical profession and the educated classes of Victorian Britain. Popular medical encyclopaedias, medical journals, and newspapers dealt with the diagnosis of heart disease, which coincided with the growth in cardiac medicine as a clinical specialism.[76] Christopher Lawrence has identified the rapid emergence of specifically qualified physicians and hospitals, and an apparent rise in heart disease as a cause of death.[77] There was concern that the rising death rate as a result of cardiac dysfunction was widespread, and that it reflected the rapid pace of modern life in Victorian Britain.

Although the 1872 *Times* article cited in Chapter 4 went on to associate heart disease (especially, I would suggest, organic disease) with working men between 21 and 40 years of age, the gendering of cardiac dysfunction was complex. Certain types of heart disease, such as angina pectoris, were skewed towards men. Yet the literary imagery of heart disease as an enfeebling and emotional disease was peculiarly female. In her work on the poetics of heart disease, Blair has convincingly claimed that literary analyses associated heart disease not only with women, but also with a peculiar sensibility and emotional susceptibility. I would suggest that this association was principally found in the diagnosis and characterization of *functional* heart diseases, as functional disease was increasingly linked to emotional excesses (and, in time, to concepts of neuroses), which were feminized during the nineteenth century.[78] In the mid-nineteenth century, however, the knowledge and possession of heart disease became a status symbol among certain female literati who combined literary acclaim with heart defects and poetic artistry. This is consistent with Blair's account of nineteenth-century writers who focused on their hearts and its movements in letters and practical experimentation, and who sought to position themselves as cardiac sufferers 'precisely because it was popularly believed to stem from acuteness of feeling, emotional sensitivity'.[79]

The Heart of Harriet Martineau

In presenting the public face of her mysterious illness as heart disease, then, might Harriet Martineau have attempted to participate in its cultural cachet, to demonstrate further the emotional sensitivity exhibited through her fictional writing and in her *Autobiography*? She would certainly have been well versed in the dual rhetoric of sensitivity and cardiac dysfunction at a time when literary and medical discussions of cardiac characteristics drew from and influenced one another.[80] On a personal level, moreover, Martineau's extensive literary and social connections connected her to several fellow sufferers *and* literary figures—one of her closest friends being Mary Arnold, Thomas Arnold's widow—for whom the experience of heart disease loomed large.[81] Perhaps even most interestingly, this network focused on the connections between these members of the literary elite and Peter Mere Latham, the foremost cardiac specialist in Victorian culture.

Thomas Arnold's death from angina pectoris in 1842 was used as a case study by Peter Mere Latham in his *Lectures on Clinical Medicine*. Thomas's son Matthew also suffered from cardiac problems, on which he referred to Latham, and Thomas Arnold and Harriet Martineau shared frequent correspondences on health concerns as well as on matters of economics and education.[82] It is likely that Latham also treated Christina Rossetti, whose probable heart disease seems to have been historiographically subsumed by her diagnosis of cancer.[83] Moreover, a disproportionate number of Martineau's female correspondents and literary associates *also* suffered from cardiac symptoms, including Elizabeth Barrett Browning, Elizabeth Gaskell (who died of a heart attack aged only 55), Mary Carpenter (educationalist and associate of Harriet's brother James Martineau), Ann Sykes Swaine (bap. 1821, d. 1883), suffragist and philanthropist, and Mary Russell Mitford (1787–1855), playwright and writer.[84]

In one exchange between Barrett Browning and Martineau, Winter records, the two intellectual women debated the relationship between creativity and nervous tension; the desire to create, or to produce (and the complex feelings that that desire produced within a suffering individual), being incomprehensible to physicians. It was pointless, Martineau observed, for them to forbid 'all excitement & intellectual labour, as if one could hush one's mind, as you pat your dog to sleep'. And, while it would be better for Barrett that her 'pulses' remained 'in order', disorder was often necessary for the creative process: Barrett would not recover while she was 'keeping a burning & thrilling weight of poetry on [the] heart and brain'.[85]

In writing about her first period of illness, Martineau had expressed the belief that her ailments were 'unquestionably the result of excessive anxiety of mind—of the extreme tension of nerves under which I had been living for some years'.[86] This observation seems to have continued to dominate Martineau's life and ill health, embodying, in the process, Victorian notions of the complex interactions between health, the circulation, and the central nervous system.[87] Indeed, Martineau's life of mental and physical suffering, as rendered in her *Autobiography*, characterized the nervous sensibility of the intellectual and creative woman.

It was entirely fitting, therefore, that Martineau should succumb to disease of the heart, an organ that was physically and symbolically at the centre of a number of ideological beliefs about emotion, gender, health, and disease. For, while Martineau was not only a writer of fiction but a social reformer and political thinker, she would also have been acutely aware of the link between the heart and creativity, and between heart disease and a refined sensibility.[88] So it was, perhaps, that Martineau chose to highlight cardiac disease rather than a uterine tumour as her self-positioning as an invalid, in much the same way as Barrett Browning and Christina Rossetti focused on their hearts rather than on consumption, or on cancer, respectively.[89]

This possibility sheds new light on Martineau's self-construction as an invalid, and in her dealings with the medical profession. Heart disease provided a way for her to discuss her ailments publicly without falling back into the gendered position of a woman who suffered from gynaecological disorders. More positively, moreover, at a time when functional heart diseases in particular—diseases of function rather than structure that were linked above all else to the operation of the nervous system—were imbued with broader discourses on mental and physical sensibility, heart disease provided a way for Martineau to rewrite her symptoms as a mark of superiority rather than debility.

The romanticization of the heart and its links with individuals' conceptions of selfhood and identity in the Victorian culture provides an interesting corollary to the disenchantment of the heart by patho-logical anatomy and experimental physiology. And its influence still lingers, perhaps because we are still influenced by romantic ideology, whether in understandings of our individual identity, or in the histori-cal evolution of scientific categories.[90] In some ways, this preservation of holistic, almost Galenic, sentiments about the heart and its centrality to human emotions within a new framework—in this case of elitist intellectual and philosophical emotional perception—helps to explain the perseverance of cultural metaphors of heartfelt affection in the modern age.

Medical interpretations of the heart were not solely influenced by cultural attitudes towards the heart, however—whether that was its status as an emotional repository or its function as a pump. More than any other factor, perhaps, the rise of the mind sciences and the emergence of secular, rational, and psychological ways of viewing the emotions helped to focus attention away from the heart and towards the brain in the understanding of emotional states. The modern self of the post-Romantic period was to be a cranio-centric, rather than a cardio-centric one.

7

Emotions and the Brain:
Rethinking the Mind–Body
Relationship

> The brain is the source of all the feelings, ideas, affections and
> passions; their manifestations, therefore, must depend on the brain
> and be modified by it.
>
> Francis Gall, *On the Functions of the Brain* (1835)[1]

Disenchantment of the heart was not enough for the brain to become the centre of emotion in any modern sense. Emotions were also linked to the organ of the brain because of interest *in* the brain, and in the complex metaphysical structure that was 'mind'. In medical and anatomical terms, the development of the brain as a symbol and organ of emotion was implicit in many of the reformulations of the brain and the nervous system that took place in the nineteenth century under the influence of experimental physiology, early psychology, philosophy, and ethics. Several historians of science and medicine have addressed the anatomical and physiological reformulations that accompanied the growth of 'scientific medicine', and it is not the purpose of this chapter to summarize them.

Nevertheless, it is important to engage, in broad strokes, with the trajectory of its development, in order that we can understand the shifting status of the mind and body, and the disjuncture that subsequently emerged between what we might term lay or ('common-sense') and scientific theorizing of the respective roles of the brain and the heart by the late nineteenth century. Here I will focus, therefore, on developments in anatomy and the mind sciences in so far as they impacted on the respective roles of the heart and the brain in the human body, and anticipated the decline of the emotional heart.

It is interesting that the competing relationship between head and heart as emotion centre has seldom been addressed as part of the story of psychology and the mind sciences. It is, perhaps, testimony to the dominance of the brain (qua mind) in Western medicine and culture that its status as the organ linked most commonly with emotion, identity, and selfhood is largely unquestioned and apparently beyond history and culture.[2] The brain has become the organ associated with the individual evolution of our selves in a post-Freudian, psychological age, and with the parallel development of humankind. And yet that association is relatively new in historical terms. Those who *do* recognize its relative newness commonly address it from a teleological perspective: how it is that the modern, bounded individual emerged—testimony to the post-Cartesian split of mind and body, as well as the impact of such socially and culturally individualistic enterprises as market capitalism and Protestantism.[3]

In unravelling the historical connections between the heart, the brain, and the mind, it is also striking how little interaction there has been between alternative ways of understanding emotional processes, such as theology and ethics, experimental physiology, and experimental psychology, all of which relied upon different epistemological frameworks to understand the meanings of brain function and the mind-body relationship. Thus, in the nineteenth century, even sophisticated accounts of physiological experiments into the function of the brain made no reference to its implications for the heart. This phenomenon has continued in the work of historians, arguably because of the disciplinary boundaries that mean that the history of neuroscience need not be connected with the history of cardiology; one is, after all, the study of the mind and the other of the body.[4] Yet the development of neuroscience must have had significant impact on scientific understandings of the role and function of the heart that cut across those emergent boundaries. In particular, the idea that the heart's operation was mechanically (and

later chemically or electrically) rather than spiritually activated, and
that it was a reactive, rather than an active, force in the production
of states such as anger and fear, was one of the most dramatic influences on redefinitions of the emotional importance of mind.
Equally significant was the problematic relationship between mind and
matter, on which centuries-old philosophical considerations continue
to turn.[5]

Materialism and the Localization of Brain Function

The relationship between mind and body, like that between brain
and mind, has been a central problematic in Western metaphysics
for several centuries. Throughout the seventeenth and eighteenth
centuries, and well into the nineteenth, concepts of holism dominated,
which meant that, despite the apparent ubiquity of Cartesian dualism
in philosophies of the mind-body relation, the two realms continued
to be interrelated and mutually interactive.[6] In the nineteenth century,
and through a series of experiments in physiology and psychology, the
'mind' was located in, and began to be synonymous with, the organ
that best represented it: the 'brain'.

There were, Robert M. Young has observed, three relevant
and materially grounded 'scientific discoveries' that were of significance in identifying mind as brain: first, the 'sensory and motor
functions of the posterior and anterior spinal nerve roots', second, the
role of the 'frontal convolution in motor aphasia', and, finally, the
role of the 'cerebral cortex in sensory and motor functions'.[7] Accompanying these reimaginings of the physicality of the brain and the
nervous system were a number of important conceptual beliefs about
brain function and about reason, memory, imagination, and other
mental processes, many of which were reduced to sensory-motor
functions.[8]

Several histories of science and medicine chart the development of
scientific understandings of the brain. One of the most comprehensive
of these is *Nineteenth-Century Origins of Neuroscientific Concepts*, in which
Edwin Clarke and L. Stephen Jacyna trace shifts in understanding the
brain and nervous system from the ideas of Galen and classical antiquity, through the romantic philosophies of the eighteenth century,
and into the die-hard experimentalism of scientists like the American

physiologist John Coll Dalton.[9] Clarke and Jacyna's material exploration ranges from studies of chemical analysis, dissection, vivisection, and comparative anatomy, to foetal analysis, pathological experimentation, and cranial comparison.[10] From the 1830s, early neuroscientists' desire to understand the physical structures of the human brain and its influence on the body and the mind developed in the same way, and at the same time, as early cardiologists' parallel desire to map the surfaces of the heart. Attempts to chart and delineate brain function were most explicit in the development of cranial localization.

In common with most writers on the history of the brain, Clarke and Jacyna identify Phrenology (initially 'Organology'), attributed to the German physician Franz Joseph Gall, as the earliest account of distinct regions of the brain being designated the seats of particular motivations and emotions. Gall's 'doctrine of the skull' reduced personality and intelligence to twenty-seven powers or functions that included 'pride', 'cunning', and 'poetic talent'.[11] Ridiculing the discussion of French physiologist Anthelme Richerand of the size of an individual's heart as a direct correspondence to the size of his or her courage, Gall gave emotional predispositions or tendencies a particular origin, though his 'affective' attributes did not focus on individual emotions, but on such general 'sentiments' as 'self-esteem' and 'benevolence'.[12] As has been detailed above, the brain was the 'undisputed organ of mind' for Gall, and it was through material means that emotions were manifested: 'the brain is the source of all the feelings, ideas, affections and passions; their manifestations, therefore, must depend on the brain and be modified by it.'[13] This work directly challenged those—like the French physician and pathologist Marie-François-Xavier Bichat—who had identified the brain as the seat of intellectual, but not emotional, life.[14]

Bichat's work is important to us in understanding both the history of the interrelationship of brain and body before the nineteenth century, and the ways principles of mutual interdependence had been ingrained in medical thought since the time of Galen. In the seventeenth century, Thomas Willis had established rules of voluntary and involuntary motion impacting on both realms (of mind and body) in his writings on the sympathetic nervous system.[15] Bichat's work merged practical experimentation with Willis's principles (and those of Robert Whytt of Edinburgh) to create an anatomico-physiological system that separated *la vie organique*, or the organic life (which included the heart, the intestines, and the other inner organs and which was connected with the passions) from *la vie animale*, or the animal life (which was concerned with the higher functions of habit and memory, will and intellect). Although the intellect was the function of the brain, it could not exist without the centre of organic life, the heart; the two organs were mutually interdependent.[16] The development of what was to become the 'vegetative' (more commonly, with the work of the British physiologists W. H. Gaskell and John Newport Langley, the 'autonomic') nervous system influenced interpretations of what Bichat termed 'organic life' by the mid-nineteenth century—the heartbeat becoming simply one of many bodily functions that was beyond 'habit' or 'education'.[17] There was no longer any need for will or judgement or the soul in the operation of the heart; it was a product of the body and separable from the more complex workings of the mind.

This separation of mind and body (and of psychological and physiological processes) was an important aspect of the emergence of materialist medicine. At the same time as the heart and its actions became redefined as simple 'reflexes', the origin of emotion was linked to the operation of the mind.[18] 'Basic passions', like anger and sadness, were redefined as mental processes that made their impact felt upon the body, disrupting the normal operation of the autonomic nervous system. In separating the brain from the heart, and by identifying the former as the physical basis of all mental functions, Gall's work was influential in maintaining this cranial emphasis, even if phrenology was otherwise short-lived and frequently satirized (Fig. 7). It influenced the work of many other 'localizers', including Sir David Ferrier, Richard Caton, and others, who worked on the electrophysiology of the brain.[19] It thus helped to identify the physical matter of the brain as the location for mental processes, and facilitated the division of mind from matter, of mental from physical phenomena, that underpinned the development of much scientific thought about the mind-body relation from the nineteenth century.[20] Perhaps even more importantly, it was essential to the reimagining of emotional experience as felt *by* the body but not being *of* the body. The true origin of emotions was re-sited in the material structure of the brain, the organ that Aristotle had famously dismissed as little more than a cooling agent for the heart.[21]

Fig. 7. Gall satire. 'Francis Joseph Gall leading a discussion on phrenology with five colleagues among his extensive collection of skulls and model heads.'
Source: Wellcome Images.

Of Mind and Matter

The mid-eighteenth-century emphasis on materialism, heavily influenced by Cartesian metaphysics and subsequently by the Romantic movement's concern for the unity of the spiritual and physical worlds, influenced the development of much biological thought well into the nineteenth century, perpetuating a tension between essence and substance.[22] Attempts to explain the origins and mysteries of human consciousness included the suggestion that it was some kind of nervous force—some material or immaterial energy—that acted as a source of power; in this, the material structure of the brain was defined as a series of 'little souls' that composed the 'general soul of the entire body'.[23] It is notable that a parallel yet materialist construction of the heart (as containing a 'little brain') was a dominant trope of one of the most controversial physiological emotion theories of the past century.[24] In much the same way as the heart could be perceived as a physical object stimulated by vital forces, so could human consciousness and thought be located in the physical, grey matter of the brain.

Of course, even if emotions were redefined as mental phenomena, the extent and nature of their influence on the body remained problematic. Mapping the physical structure of the nervous system and identifying its nervous pathways and ganglions were relatively straightforward. So, too, was theoretically separating the functions of the central nervous system (the brain and spinal cord) from those of the autonomic nervous system. Far more difficult was explaining the various ways in which the systems were interrelated. Understanding the qualities that brought the brain to life, that animated thought, feelings, and emotions, was a metaphysical and philosophical, as well as a scientific and material problem. It was also the same problem as theorists of the heart had wrestled with in their discussions of the enervation of the heartbeat and the actions of the pulse.[25]

Moreover, if mental processes were to be relocated in the brain, what kinds of relationships between mind and matter were at stake? How could feelings such as love, anger, joy and hope be converted into physical activities, material movements along a nervous thread? More fundamentally, how could emotions originate in the brain if not through the influence of the soul? Was there some essential and separable aspect of 'mind' at work, or were the mind, and the feelings, memories, and emotions, merely physical, concrete entities that could be cognitively processed? The existence of such questions meant that, perhaps unsurprisingly, not all theorists believed in the reduction of emotion to physical processes, or the reduction of mind to matter. For many nineteenth-century researchers, it was not whether immaterial and psychological influences made themselves felt on the material and physiological realms, but *how* they did so.[26] The role of cognition in this process was crucial.

Cognitive Theories of Emotion

Under a cranio-centric model of emotions, human affects and motivations were somehow linked to the brain's cognitive and structural features. Of course, the body might *feel* emotions (and those emotions might be manifested through a raised heartbeat or a cold sweat), but those emotions were cognitively processed by the brain as a thinking and feeling organ in much the same way that they were once believed to be processed by the soul. Mediated by or without an immaterial concept of 'mind', the role of the brain as the cognitive centre (the term 'cognition' from the Latin 'to know' or 'to recognize') was integral to the rise of the mind sciences.

Despite its relevance to the processes of cognition, however, emotion was not explicitly addressed in most physiological researches into the brain from the eighteenth century. From the Edinburgh-based work of Alexander Walter and Charles Bell to François Magendie's experimentations in the spinal roots of dogs, most research was focused on sensation and movement (and later, with the work of Paul Broca and Carl Wernicke, with language).[27] This was one of the complaints made by the American physiologist William James, in his now classic 'What is an Emotion?' (1884). Highlighting the importance of the body's influence on the brain as a way of understanding emotion physiology, James rejected most physiological experimentation, and those physiologists who

during the past few years, have been so industriously exploring the functions of the brain, have limited their attempts at explanation to its cognitive and volitional performances. Dividing the brain into sensorial and motor centres, they have found their division to be exactly paralleled by the analysis made by empirical psychology, of the perceptive and volitional parts of the mind into their simplest elements. But the aesthetic sphere of the mind, its longings, its pleasures and pains, and its emotions, have been so ignored in all these researches.[28]

James's best-known article on the subject proposed that the nervous system was predisposed to respond in particular ways to the environment. In essence, he argued it was physical sensations, rather than cognitive processes, that triggered an individual's bodily experiences of emotion, from a raised heartbeat to a cold sweat. Only after those visceral changes did the brain respond by conceptualizing an emotion as fear.

James and Cannon: Opposing Sides of the Mind–Body Argument

James therefore challenged the polarized thinking about the brain that had dominated discussions of localization since the work of Gall. Either emotions resided in the brain, he wrote, under 'separate and special centres' (a theory challenged by work into frontal-lobe function), or they 'correspond to processes occurring in the motor and sensory centres', known or otherwise. If the former was true, then conceptions of the brain needed to be more sophisticated; if the latter, then psychologists needed to consider whether the processes were peculiar, or similar to other 'perceptive processes'.[29]

James's 1884 essay took up the latter position, arguing that emotional brain processes were not separate from, but a combination of, other brain processes. Understanding this, he maintained, would simplify brain physiology, and help theorists to understand emotions themselves.[30] Basing his assessment only on emotions with 'a distinct bodily expression' (and not on those that brought some indistinct sense of satisfaction or intellectual gratification), James included 'surprise, curiosity, rapture, fear, anger, lust, greed, and the like' as 'mental states with which the person is possessed.' Our 'natural way' of thinking of these 'standard emotions', wrote James, was that the mental perception of a fact excites the bodily expression. In other words, one became frightened and shook with fear. James turned this theory on its head, claiming that the perception of an event brought bodily changes first (the shaking with fear), and that those bodily changes, with our awareness of them, composed the emotion. Thus:

Common sense says, we lose our fortune, are sorry and weep; we meet a bear, are frightened and run; we are insulted by a rival, are angry and strike. The hypothesis here to be defended says that this order of sequence is incorrect, that the one mental state is not immediately induced by the other, that the bodily manifestations must first be interposed between, and that the more rational statement is that we feel sorry because we cry, angry because we strike, afraid because we tremble, and not that we cry, strike, or tremble, because we are sorry, angry, or fearful, as the case may be. Without the bodily states following on the perception, the latter would be purely cognitive in form, pale, colourless, destitute of emotional warmth. We might then see the bear, and judge it best to run, receive the insult and deem it right to strike, but we could not actually feel afraid or angry.[31]

James's inversion of the 'common-sense' model was based on the generalized belief that the nervous system was a mass of predispositions, and the nerves merely 'a hyphen between determinate arrangements of matter outside the body and determinate impulses' within it. The emotions formed part of the 'nervous anticipations' that spurred the body's impulses, and these were predetermined. For what child could see an elephant thundering towards it without experiencing fear, even if he or she had no knowledge of what the beast was? Similarly, women 'feel' delight at young children even without consciously recognizing that delight.[32]

These were our inherited emotional propensities, James explained, in a way that was resonant with Darwinian theory. Indeed, he explicitly referred to Darwin's *Expression of the Emotions*, and to Charles Bell's *Anatomy of Expression*, as evidence of the consistent and external visible symptoms of emotion. James also referred to Angelo Mosso's plethysmograph, one of the instruments that transformed the physiological measurement of emotional experiences in the late nineteenth century, as evidence of the uniformly measurable quality of bodily responses.[33] This work, James claimed, showed how the nervous system made the brain's movements known to the whole of the body and its organs, including the heart. Moreover, he added:

not only the heart, but the entire circulatory system, forms a sort of sounding board, which every change of our consciousness, however slight, may make reverberate. Hardly a sensation comes to us without sending waves of alternate constriction and dilatation down the arteries of our arms. The blood-vessels of the abdomen act reciprocally with those of the more outward parts. The bladder and bowels, the glands of the mouth, throat, and skin, and the liver, are known to be affected gravely in certain severe emotions, and are unquestionably affected transiently when the emotions are of a lighter sort. That the heartbeats and the rhythm of breathing play a leading part in all emotions whatsoever is a matter too notorious for proof. And what is really equally prominent, but less likely to be admitted until special attention is drawn to the fact, is the continuous co-operation of the voluntary muscles in our emotional states.[34]

James's explanation of emotion physiology effectively reduced emotions to their physical effects, for 'what kind of an emotion of fear would be left, if the feelings neither of quickened heart-beats nor of shallow breathing, neither of trembling lips nor of weakened limbs, neither of goose-flesh nor of visceral stirrings, were present, it is quite impossible to think'.[35]

An alternative approach to understanding emotions can be found in the work of the American physiologist Walter Bradford Cannon.[36] After analysing emotions in various organ systems and functions Cannon viewed the brain and the brain *alone* as the centre of all activity, emotional or physical. Contrary to the claims of James, he believed that emotional responses were first cognitively processed by the brain and then relayed to the body, to the cardiovascular system and the heart.[37] In Cannon's now classic 1929 publication of *Bodily Changes in Pain, Hunger, Fear, and Rage*, emotions were studied alongside other bodily processes and systems as fundamental physiological events.[38] What is interesting about this work in relation to the heart and to emotions is that Cannon incidentally raised the question of the extent to which changes in the heartbeat described changes in 'feeling' at a subjective level.[39] In short, he highlighted the problems in terminologies of emotion, as well of the potential gap between physical evidence and cognitive experience.

The recognition that the heartbeat was not *always* dependent on cognitive processes, and that it could, instead, be dependent on visceral changes, including those produced by hormones, meant that 'apparent, visible emotion does not necessarily correspond to the patient's real feelings at the time'.[40] The heartbeat and its origin had long been the subject of experimental research and observation, as characterized by the Cambridge school of Physiology under Michael Foster.[41] As a process, however, and as a product of the autonomic nervous system, Cannon's work showed that the heartbeat could become both rationalized and explicable as a result of nervous processes regulated by the brain—no more and no less symbolic in respect of the emotions than such parallel physiological events as sweating and blushing.

An Evolutionary Perspective

The work of both James and Cannon stressed the influence of Darwinian evolutionary theory. Cannon's work explicitly followed the trajectory of Darwin's, by presenting an evolutionary model of the nervous system. Indeed, this evolutionary model dominated much neuro-physiological work in the twentieth century, a point succinctly argued by Rhodri Hayward in his study of William Grey Walter.[42] Darwin's *Expression of the Emotions in Man and Animals* stressed the impact and origins of emotion indicators in ways that helped to explain how bodily expressions of emotion—including an increased heartbeat—might be linked to cognitive processes that underpinned human nature.[43] In the case of rage, for instance, Darwin wrote that 'the action of the heart is much accelerated, or it may be much disturbed' by a slight to the self. 'The face reddens, or it becomes purple from the impeded return of the blood, or may turn deadly pale.'[44]

This description of the visible effects of emotion is reminiscent of those found in seventeenth-century accounts of anger: the face reddening, the teeth clenching, and the eyes shining with fury.[45] And yet there is an important distinction. Rather than being an active agent in those bodily changes—the heart and the soul working together to summon spirits and bloods in order to provide an appropriate emotional response—the heart is a recipient of the mental perception of an injury—threatened or real—to the person. It is the nervous system that becomes part of an organizing principle of evolution, rather than the heart. The influence of the heart is simply a mechanical one derived from a system of cause and effect: disturbed by an overactive nervous system, it begins to move the blood more forcefully around the body, or ceases to perform its circulatory function correctly, impeding the return of blood to a face that becomes 'deadly pale'.

In the 1880s and 1890s, the English experimental physiologist John Hughlings Jackson adapted Darwinian evolutionary theory to build a conception of brain function as a series of layers, each of which was progressively more advanced than that which preceded it.[46] In ways that remain influential today (largely as a result of Sigmund Freud's adaptation of the schema in his identification of the *id,* the *ego,* and the *superego*), the emotions became associated with the most primitive functions of the brain.[47] Certain emotions, like anger, were believed to be more 'primitive' than others, most notably love.[48] Interestingly, this patterning of the relative primitivism of certain emotions was consistent with the evolution of philosophical theory since the eighteenth century, with its prioritization of 'social' over 'selfish' passions as the archetypal features of modern, civilized societies.[49]

By the end of the nineteenth century, the inverted traditional hierarchy of the brain and the heart in emotion theories was reinforced by experimental physiology and the emergence of a 'science of emotion'. As outlined above, Otniel Dror has shown that there was a trend towards establishing categories of emotions and their effects in nineteenth- and twentieth-century scientific culture.[50] This was arguably a parallel to the processes of investigation and experimentation found in early cardiology. And yet researchers into emotions were, in the main, uninterested in cardiac function except as a way of understanding physiological processes. The philosophical and spiritual impact of the heart, and its links to emotion causation, were entirely removed.[51]

Scientific quantification of emotion brought its own problems, however, not least through the problematic nature of terminology, already alluded to above, and through the potential for disjuncture between experience and expression. The reduction of the heartbeat to an autonomic impulse did not, therefore, end debates about the relationship between emotion and feeling. If the heartbeat could simply increase as a response to physiological changes in the body, including an influx of emotions, how could it be any indicator of an individual's emotional state? Put brutally, how could physiologists identify shifts in the measurement of pulse rate or adrenalin through the body as indicative of psychological shifts? How could 'emotion', rather than its correlates, actually ever be measured?

In 1897, the English edition of the German psychologist Wilhelm Max Wundt's *Outlines of Psychology* (translated by Charles Hubbard Judd) began to address these kinds of questions. Wundt's work discussed emotions as transient states that influenced the whole body as a result of their stimulant quality:

There is not only an intensification of the effect on heart, blood-vessels, and respiration, but the external muscles are always affected in an unmistakable manner . . . In the case of stronger emotions there may be still more extensive disturbances of innervation, such as trembling, convulsive contractions of the diaphragm and of the facial muscles, and paralytic relaxation of the muscles.[52]

Interested in how far physiological expressions of emotion were reflective of psychological experiences, Wundt and his research team performed detailed experiments at Leipzig University to measure 'conscious processes' and 'immediate experience'.[53] Relevant concerns included the quantification of emotion activity, as well as less technological tests of reaction times and word associations.[54]

The changes in measurable cardiac activity that accompanied emotions were often minimal, reflecting only slight measurable shifts when plotted by graph, or revealed in 'retarded and strengthened pulse-beats, since the intense excitation affects most the inhibitory nerves of the heart'.[55] In describing the physical effects of emotions on the organs of the body, especially in the 'pulse, respiratory organs and blood-vessels', Wundt concluded that it was possible for a variety of similar expressive phenomena to be present in diverse emotional states. It was also possible for 'feelings' to be present that might have nothing to do with 'emotions'. As Wundt saw it, the difference between the two was 'psychological' (perhaps meaning cognitive), in nature:

The emotion is made up of a series of feelings united into a unitary whole. Expressive movements are the results, on the physical side, of the increase which the preceding parts of such a series have on those succeeding. It follows directly that the deciding characteristics for the classification of emotions must be psychological.[56]

The identification of the autonomic nervous system, and the discussion of the heartbeat as an involuntary action, apparently no more symbolic than 'gastric secretion', then, arguably made assessments of 'real' feelings and 'true' emotional experiences more contested than ever before.

Emotions, the Self, and Modernity

The kinds of questions posed by Wundt are essentially those that continue to polarize opinion amongst emotion theorists today. In measuring or describing emotions, how can we know that we are talking about feelings, rather than the physiological or psychological manifestations of feelings? Or, more contentiously (and perhaps more authentically), what if feelings are all that there are, products of culture that are mapped onto (or constitutive of) experience in different ways and at different times. There is insufficient space here to discuss the changes taking place in physiologies, psychologies, and even philosophies of emotion in the twentieth century. The sheer proliferation of theories grounded in biology, psychology, neurobiology, neuropsychology, sociology, and anthropology is too vast to be considered here.[57] What should be noted, however, is that, throughout all such approaches, and with few exceptions, the principle of brain-based emotion has predominated. There has been comparatively little research into emotions as products of the body (though there has been an emphasis on endocrinology, for instance, as a way of explaining the interaction of the mind and body in material terms).[58] In non-Western and non-orthodox medical traditions, moreover, the heart continues to dominate as an explanatory category in Eastern and complementary practices.[59]

The historical development of physiological psychology, and its links to debates in experimental physiology and evolutionary theory, show how complex the meanings of the emotional brain and its links with the heart were by the nineteenth century. In etymological terms, 'psychology' was once conceived of as the 'study of the soul', as the soul moved in and through the mind to influence the perception and generation of emotional states, the action of the heartbeat, the measurement of the pulse.[60] By the mid-nineteenth century, the emphasis had shifted from mind to brain, but the brain was not immediately secularized. The centrality of the soul as a persistent force in early psychological theory continued to exert influence, underpinned as it was by many of the assumptions of natural theology.[61]

The relatively slow decline of the soul in psychological and scientific theory demonstrates its strength and versatility as an explanatory category. Without it, any articulation of the mental and physical processes were dogged by the same issues common to metaphysics and medical sciences alike since humoralism: the relationship between mind and body, or between mental cognition and physical experience, and how those two entities could be conceptualized in one comprehensive and measurable framework.[62] The most dominant model presented here was that of the brain and the central nervous system.

But scientific debates in the 1800s were not unilaterally in favour of viewing the brain as the unitary object of emotion, any more than they were separable from religious debates about the relationship between mind, brain, and soul. The influence of the soul arguably did not diminish until the 1890s, with the arrival of modern experimental psychology.[63] Moreover, as Thomas Dixon has argued in his work on the transition from 'passions' to 'emotions', the emergence of a dominant scientific and secular model of emotion causation should not blind us to the fact that theological or metaphysical explanations continue to exist; they are simply not granted intellectual or academic status in the West.[64]

Despite such continuity, scientific interpretations of emotions undoubtedly impacted on the way they were perceived as physical and psychological as well as philosophical entities. The measurement of mind and body interaction, through the elaborate systematization of techniques to understand the heartbeat or the myriad physiological changes accompanying emotion, arguably made emotions more 'private' at the same time as they became more quantifiable and 'public' events. This was encouraged by the development of psychology and the reduction—most notably by Freud—of the individual to a unitary, monadic entity—one's circumscribed individual experience defining the course of one's life.[65]

This does not mean that the identification of the brain rather than the heart made emotions any more or any less 'social' events than they always had been. But it does mean that the accompanying medical and cultural associations of the brain and the heart as organs—the latter becoming separable from the rest of the body, a pump, responsible for the automatic processes that kept the body functioning, and the former redefined as the site of memory, cognition, and feeling—identified the brain rather than the heart as the repository of all those secret, individual thoughts and predispositions believed to make us human. When combined with the development of theories of character and personality from the late nineteenth century, and with the habitual identification, in the West, of Cartesian theory as a way of understanding the relationship between the mind and the body (the mind being superior, rational, male, and logical; the body as inferior, unregulated, irrational, female, and illogical), it is easy to understand how it is that the brain has come to be the organ *par excellence* of modern conceptions of interiority and selfhood.[66]

It is interesting that this privileging of the brain as an organ of the self is reinforced by the early twentieth-century status of categories of medical knowledge: we can change our hearts when they are worn out, broken, or damaged (albeit with a variety of philosophical and spiritual, as well as practical, challenges), but we cannot change our brains. The idea of a head transplant is anathema (in much the same way as heart transplants once were), popularly associated with body snatching and identity theft, and invoking comparisons with Frankenstein's monster.[67] The brain has become embedded in our notions of selfhood, memory, imagination, and the emotions. Yet, as testified by the predominance of hearts rather than brains on Valentine's Day cards, the organ lacks the aesthetic and emotional resonance of the heart. As a symbol of feeling and emotion, the heart is unbeaten.

Conclusion: The Matter
of the Heart

I laid it back quickly against the wall, and a sort of hot something
began swelling outward from a point in the centre of my breast;
whenever I look at a great picture for the first time I know why we
still speak of the heart as the seat of emotions.

John Banville, *The Untouchable* (1997)[1]

This book has looked at two different hearts: the heart of emotion
(also known as the heart of 'romance', or the heart of 'feeling'),
which stands as the informally acknowledged centre of embedded
cultural beliefs, and the heart of science, the heart that can be
measured, compared, treated, replenished, or replaced. Invocation
of each of these hearts conjures up its own set of assumptions.
The differences and similarities between the hearts are physical as
well as symbolic in their representation—consider, for instance, the
starkly coloured blues and reds of the student's anatomical model set
alongside the perfectly symmetrical and tongue-pink or blood-red
hearts of St Valentine (Figs. 8 and 9).

Locating the heart of the matter in the matter of the heart is
far from straightforward. One of the most remarkable aspects of
the medical and cultural history of the heart is the longevity of
its emotional meanings. There are many reasons for this—religious,
symbolic, linguistic, philosophical, and medical. What this book has
attempted to do is to show how those meanings were understood
and processed within medical and cultural discourses about the heart,
as related to, and distinct from, emotions and the brain. Aside from

Fig. 8. Valentine heart.
Valentine's Day card (1928), showing the traditional representation of the heart as a symbol of love.
Source: Wellcome Images.

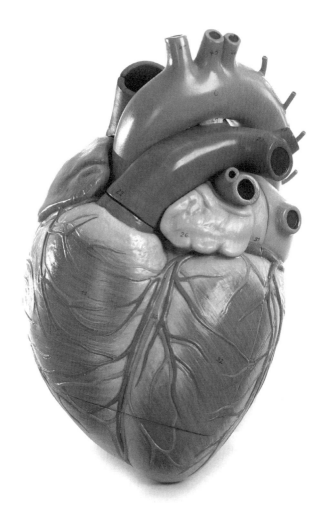

Fig. 9. Anatomical heart.
Anatomical heart (2004) used for teaching purposes, the blue superior vena cava existing from the right atrium, and the large red aorta which originates from the muscular left ventricle.
Source: Wellcome Images.

this comparative, historical agenda, a related concern is the impact of those historical meanings on the ways we view the heart today.

The broad sweep through time taken by this book—from the seventeenth to the twentieth century—is ambitious, perhaps recklessly so. Yet to understand the history of concepts, and of the changing status of emotions *as cultural artefacts*, rather than as stable constants, it is precisely this kind of *longue durée* approach that is required. There is no linear narrative constructed by this project. Rather, we have a series of sections or slices of the heart, served up at various points throughout the book in order to locate or analyse its meanings.

In addition to summarizing the main points sketched out above, this conclusion broadens the terms of debate by exploring further our deeply ambiguous relationship towards the heart of emotion in the Western medical tradition. In order to concretize some of the points it has raised, it will end with the case of an imaginary, though not entirely fictitious, patient, a composite constructed from the case studies of physicians and medical theorists, all of which had basis in some carefully identified individual experience.

In terms of the medical world of the seventeenth century, our patient, a 48-year-old woman—let us call her Elizabeth—visited her physician to complain of physical symptoms of anxiety after a disagreement with a family member. Those symptoms could have included pain at her chest and a chronic shaking in her limbs. She might also have been constipated and experiencing palpitations at the heart, along with the sensation that 'a bird was trapped inside her chest and struggling to get out'. Relatives and friends of Elizabeth might have described her as a choleric and angry woman, habitually prone to confrontations with her friends and family. It would have been unsurprising, then, that what contemporaries perceived as 'bad blood and bitterness' could have been trapped like poison in her body and affected her heart. This condition made perfect sense in early modern physiology. Elizabeth's ailment would have been exacerbated by her constipation, and the fact that she was post-menopausal. Without such outlets as stool excretion and menstruation, the 'hot blood' that was stuck in her body, brought on by and maintaining her anger, could have warmed her heart and possibly endangered her life.

Any physician worth his salt would have known just what to do to release those noxious vapours: an emetic might have been prescribed to make Elizabeth vomit. A laxative, rhubarb tincture, could have been recommended to clear out her intestine. And, finally, a little light bleeding should have removed any excess choleric blood within the system and rebalanced her mind and body. In these ways the emotions that battered away at Elizabeth's breast and gnawed away at her heart could be eradicated. Within a week, she could expect to feel calm and content. Her heart palpitations would have been eased, and she would have rediscovered her regular bowel habits. Equilibrium would be restored—at least until the next time that Elizabeth was involved in a family dispute.

Many of these assumptions made by early modern patients and physicians seem alien to us today. They envisage the heart as the centre of an individual's emotional and psychological life, as well as the site of much physical activity. They relate such sensations as palpitations to the quality and quantity of the fluids and spirits that moved between the two, connected, realms of body and mind. And they portray emotions almost as tangible, solid things, trapped in the blood and the bile and the heart of individuals, creating blockages, obstructions, and floods until physically and forcibly removed by bleeding, by emetics, by laxatives.

This interpretation of the emotional body was commonplace. As the physician William Clark put it in his *A Medical Dissertation Concerning the Effects of the Passions on Human Bodies*, it was only by 'due equilibrium' between the fluids and the solids (the organs) that individuals remained healthy.[2] Emotional extremes like anger and fear tipped the balance. Specific emotions caused specific illnesses; fevers often followed on from anger and sorrow. Epilepsy was caused by fear; melancholia by extremes of sorrow and joy. The heart was one of the most important organs, or 'solids' of the human body, and therefore at the centre of the physiological and psychological changes wrought by emotions. Medical treatises recorded how often hearts grieved and sank (often fatally) by being uplifted and defeated by their owner's emotional lives. The movement of the heart was also responsible for inciting passions, for creating bursts of joy that zipped hot blood around the body, flooding the organs, enlarging the arteries, at such speed and with such intensity that an individual could literally die from pleasure.

We cannot understand the heart's emotional status as a matter of fact and physiology by removing the heart from the breast and holding it up to scrutiny. Rather, it needs to be understood *in situ*, as an aspect of the human body, and its relationship with the emotions, the mind, and the soul. As seen above, the heart was at the centre of the emotional and physical worlds for seventeenth-century men and women, and directly related to the physiological processes that took place under humoralism. Under humoral physiology, Elizabeth's choleric nature would have made her prone to anger, just as an excess of melancholy might have made her prone to sadness.

Elizabeth's emotional state would have been dependent not only on the constitution of her body, on her gender or her lifestyle, however, but on the way her soul responded to provocation. There was a cognitive or value-judgement involved in emotions, because the heart could be 'offended' and stirred to action. This provides our best clue to the relationship between the heart and the soul, a theme central to the symbolism of the heart in religion, art, and literature for many centuries. In humoral physiology, the body and the soul were indivisible. There was no doubt that it existed, for without it how would there be life, movement, or comprehension? It was the soul that mediated between human action and divine intervention, and the soul—acting in and through the human mind—that determined the appropriate, moral, response to any given stimuli. Even after the decline of humoralism, and the soul, emotions continue to be accorded a cognitive dimension. In terms of evolutionary psychology, for instance, emotions are our developed responses to the external world and its stimulae, and necessary responses for survival.

Moreover, many of the emotional expressions we continue to use—that describe individuals as 'cold' hearted or 'hot' blooded, and experiences as 'heart-warming', 'heartening', or even 'heartfelt'—are remnants of that very real system of humoral medicine that remained intact for two thousand years. In the early twenty-first century such phrases linger mainly as linguistic habits. The heart is not commonly regarded as an active organ that heats the body and controls its circulatory fluids. It is most frequently referred to metaphorically as a pump, the vessels it supplies as pipes. In modern medical terms, the action of the heart is determined not by the operation of the soul, of course, but by electrical impulses. Its links with the emotions, with the blood vessels, and all the body's organs are explained through the actions of the autonomic nervous system that conveys impulses to the brain. Moreover, it is the brain that has become the centre of our emotions and the essence of ourselves. The heart is an object that can be transplanted, repaired, and reproduced, its functions detectable to the naked eye. This understanding of the heart has become the 'official' medical version: one school exercise used today shows children how to assemble their own 'hearts' from basic materials: a wide-mouthed plastic jar, a balloon, a skewer, and a couple of straws.[3]

This historical shift in viewing the heart in science has largely taken place since the late nineteenth century. As seen above, it was comparatively recently that the medicalized heart was detached from its spiritual and emotional influence and regarded as a material organ, subject like any other to disease and decay. This transition has not been without its controversies. For some people there remains an element of the heart that retains such emotional and spiritual meanings. Consider, for instance, the families of heart-transplant patients who claim that some aspects of personality, of temperament, or memory were transplanted along with a donor heart into the bodies of their loved ones. They are not convinced by materialist descriptions of hearts as replaceable, removable objects.[4]

One possible reason for these controversies, and why the heart has continued to beat as something more than a mere organ—indeed, as an emotional repository—is that Western scientific practice does not provide the all-encompassing understanding of the body as system that was available under humoralism. In other words, the holism of humoralism—its ability to join up the soul, the brain, the emotions, and the heart—provided some kind of explanatory power that is missing in modern medicine. In part, this is because of a focus on medical specialization that has made the heart an organ of the body, the brain the source of our selves, and the soul as redundant.

Let us imagine, in necessarily exaggerated terms, how a GP of today might respond to the complaints of Elizabeth, our patient of 1700. To recap, her symptoms—that took place after a quarrel encounter with a neighbour—included palpitations and constipation. She was known to be a choleric or 'angry' woman, a temperamental description that would still make sense to a physician. Her post-menopausal status would provide important contextualization. The first thing that the physician would do is separate her emotional and physical symptoms, her heart from her bowels, from her psyche. Less emphasis would be placed on Elizabeth's subjective awareness of her own heart function than what could be determined by a series of objective tests. The first instrument used by the physician would be the stethoscope, to listen for abnormalities in heart function. An ECG or electrocardiogram might also be recommended, for which the patient would need to travel to a specialized cardiological unit. A series of generic blood tests might be taken to eliminate thyroid and liver dysfunction, and a full blood count would be taken.

Once any material cause for Elizabeth's symptoms had been ruled out, the physician might address her emotional condition. Since she was post-menopausal, he or she might suggest hormone replacement therapy to deal with her 'mood swings', though without any other symptoms she would be unlikely to need a referral to a gynaecologist. More than likely, Elizabeth's anger would be attributed to 'stress' (a term that came into being in the early twentieth century and has become ubiquitous). She might be offered counselling or referred to a psychologist, someone qualified to deal with emotions as products of mind. And her constipation? If particularly unusual or prolonged, this might require a visit to a gastroenterologist.

In modern-day medical practice, then, the heart is not the centre of the emotional and physical body, but an organ that requires its own specialism: cardiology. The body itself is split into sections: the circulatory or cardiovascular system, the nervous systems, the digestive system, the endocrine system, the immune system, and so on. Many of the ways that the mind and body are joined up, in the absence of a soul, is related to the sensory impulses conveyed from the heart and all the body's organs through the nerves to the brain. The felt impact in the heart of certain emotional changes is attributed to hormones (including 'happy' hormones like endorphins. Is this so different, in principle, I wonder, to the concept of humours?[5]) Importantly, the heart's spiritual meanings have all but disappeared in scientific terms: the notion of the impulse or reflex has replaced the soul's role as stimulus and mediator between mind and body, head and heart.

Many of these aspects of modern medical practice—a focus on diseases rather than the individual, on objective measurements rather than the patient's interpretation of symptoms, and on specific organs rather than the system as a whole—did not come into being until the late nineteenth century. The same is true of attitudes towards the heart, and its links with emotion. It was not until there were viable, secular models of the origin of knowledge that it was possible to imagine emotions as separate from the heart, or the body as separate from the mind. Well into the nineteenth century, moreover, many medical beliefs and practices continued those that were so popular in Elizabeth's time: an emphasis on the humours, on bloodletting, on patients' descriptions of illness, and on illness itself as humoral imbalance, rather than the result of external, pathogenic influence.

It was only with the emergence of scientific medicine in Britain and continental Europe that we find the first real shifts in understanding the heart as an object of theory, reinforced and made possible by a form of medical specialization that was practised in pathological laboratories and conducted on hospital wards.[6] Through the popularization of such technologies as the stethoscope, it became increasingly possible for physicians to gain objective measurement of the action of the heart, and to detect anomalies according to a newly determined set of 'normal' measurements.

It used to be the case that only through autopsy could the failings of the heart be understood. Yet, under the influence of French physicians such as Theophile Laennec, inventor of the stethoscope, and Jean-Nicholas Corvisart, personal physician to Napoleon Boneparte, the heart's diseases could be described, classified, and compared while the patient was still alive.[7] Cardiology's authority and importance as a medical specialization were shown by the building of several heart hospitals, some of which, like the Royal Brompton, originated as centres for Consumption and Diseases of the Chest. These architectural structures were symbolic as much as functional: they made statements about physicians' ability to know living hearts as well as their eighteenth-century equivalents had known those of the dead.

At the same time as this focus on the heart as a physical organ eased away its spiritual and emotional vestments, the birth of the mind sciences—of psychology and psychiatry—emphasized that it was the brain (not the heart) that was the centre of life and of emotions. From Charles Bell's *Idea of a New Anatomy of the Brain* (1811) to Magendie's *A Summary of Physiology* (1822), neuro-physiological investigation focused on the reflex and the function of the central nervous system.[8] In clinical experiments, the heart was seen to respond to certain emotional states—as demonstrated by palpitations in fright, or by a fast heart rate in anger—but these were understood to reflect nervous or electrical communication between the brain and the nervous system, on the one hand, and the heart and cardiovascular system, on the other. Crucially, they did not denote any active emotional qualities or experiences within the heart itself.

As has been shown above, scientific medicine originally derived its status from *taking apart* the human body, from viewing it as a series of interactive but separate systems, rather than an integral whole. Not all medical practitioners were satisfied with that division, any more than they are today. Nineteenth-century treatises on heart disease show a discomfort with the distinction between mind and body, brain and heart, that this new thinking imposed. The physician Byrom Bramwell, pathologist at the Edinburgh Royal Infirmary, wrote in 1884 that it was impossible to determine the exact relationship between emotions and the heart.[9] He was content to believe that there were issues that were beyond the remit of the physician; that aspects of some divine, supernatural, or spiritual force could not be quantified or uncovered.

Evidence from mid-nineteenth-century casebooks, moreover, shows that, while practitioners used technologies like the stethoscope to measure the beating of the heart, they also listened to the patients' accounts of their overall health and well-being. They asked questions about the environment and constitution of the patient, about their diets and exercise habits, as well as the food and drink consumed. Many of these concerns are reminiscent of the 'non-naturals' that dominated humoral beliefs about the emotions and the maintenance of health. From time to time physicians even recommended bloodletting as a means to remove excess blood around the heart, or to remove imagined and unidentified blockages, and therefore to eradicate the symptoms associated with anxiety, nervousness, and depression.

Theoretically, we live in a secular and scientific world, a world shaped in part by the transformations of knowledge that emerged in the nineteenth century. Certainly many of the disciplinary categories and divisions that were developed during that period, including cardiology, remain in place. And yet we are *still* trying to make connections between these various realms of experience, to ask how it is that the relationship between emotions and the heart remains problematic. And so it is that we have ended up with two different hearts: the heart of science that is a pump, versus the heart of culture, the 'common-sense' idea that grounds feeling in the organ and the symbol of the heart.

In the twenty-first century, we still ask questions about the kinds of structures—material or otherwise—that can explain the nature of emotions, and the continued symbolism of the heart as a way of identifying ourselves. Some of the most cutting-edge developments taking place today by medical researchers break down divisions and explore some of the interconnections between mind and body. Perhaps the most dramatic example of such research is that which seeks to demonstrate the consciousness of the heart, and of the heart acting in conjunction with the body.[10] Rather than viewing the heart as a pump, for instance, Andrew Armour and others have famously reconceptualized it as a highly complex, self-organizing processing centre with its own functional 'brain' communicating with and influencing the cranial brain via the nervous system and hormone pathways.[11] This rethinking of modern scientific ideas about the heart and the mind–body relationship provides a more holistic version of the body than was previously possible. It is a scientific version of humoral, joined-up thinking about mind and body, and it fulfils a *social* as well as a purely medical function. That such research is being conducted using scientific methodologies and presented in scientific language (of 'processing centres' and 'hormone pathways') is also evidence of the medical profession's need for an alternative to the 'unscientific' and 'popular' arguments made by non-orthodox healers and practitioners.

However bizarre humoral medicine seems today, then, with its fluid economy of melancholy, choler, blood, and phlegm, its balance of the humours as a way to achieve mental, physical, and spiritual equilibrium, and its understanding of the heart (rather than the brain) as the centre of our emotional intelligence, it lasted more than two thousand years as a system of belief. Perhaps part of the reason for that is that it was able to accommodate mental and physical phenomena, thoughts, feelings, desires, within a complementary and explanatory framework that was accessible to all. In so doing, it placed the heart within a system of mind, body, and spirit that modern medical practices—through ideological, logistical, or financial constraints—do not allow.

In concluding this work, I am not privileging the status of early modern medicine over modern scientific practice. Nor am I suggesting that there is something spiritual or mystical about the heart that has yet to be 'discovered' by science. What I am suggesting is that, as an explanatory category, science has failed to conceptualize the meanings attached to the heart in ways that could make irrelevant, or replace, traditional beliefs about its emotional embodiment. The reason why this is so is that it has been constructed in opposition to, rather than in integration with, a holistic interpretation of the body.

This need not be the case. There are many other medical models that continue to celebrate holism. An interchange between emotions and the body remains at the very heart of Ayurvedic and traditional Chinese medicines, for instance, just as it was once at the centre of Western medicine.[12] Each of them also focuses on maintaining health—through respiration, circulation, digestion, elimination, exercise, and so on—in ways resonant of classical medicine and the 'non-naturals'. Moreover, some medical specialists are shaping their diagnostic and prognostic categories in ways that allow for interplay between psychical and physical factors, and view 'the patient' in the context of his or her social and emotional worlds. The work of Professor Martin Cowie, Consultant Cardiologist at the Royal Brompton Hospital, for instance, has explicitly drawn parallels between grief and reduced life expectancies, suggesting that in the first six months after bereavement a bereaved partner has a higher risk of death.[13]

It will be interesting to see how scientific research continues to engage with, or counter, alternative theories of health and disease, and perhaps rejoin these gaps between mind and body, brain and heart, in forthcoming decades. As the now well-established and scientifically legitimized divisions of the mind and body (and its carefully delineated systems and process) begin to erode, we may have more in common with Elizabeth, and the world of 1700, than we think.

Notes

INTRODUCTION

1. Ali Mazrui, 'The Poetics of a Transplanted Heart', *Transition*, 35 (1968), 51–9, 56.
2. Mazrui, 'Poetics', 51, 53, 56. On the politics and problems of organ transplantation, see Renée C. Fox and Judith P. Swazey, *Spare Parts: Organ Replacement in American Society* (Oxford: Oxford University Press, 1992); Susan E. Lederer, *Flesh and Blood: Organ Transplantation and Transfusion in Twentieth-Century America* (Oxford: Oxford University Press, 2008); Margaret Lock, *Twice Dead: Organ Transplants and the Reinvention of Death* (Berkeley and Los Angeles: University of California Press, 2002); Lesley A. Sharpe, *Strange Harvest: Organ Transplants, Denatured Bodies, and the Transformed Self* (Berkeley and Los Angeles: University of California Press, 2006); ead., *Bodies, Commodities and Biotechnologies: Death, Mourning and Scientific Desire in the Realm of Human Organ Transfer* (New York: Columbia University Press, 2007); Catherine Waldby and Robert Mitchell, *Tissue Economies: Blood, Organs and Cell Lines in Late Capitalism* (Durham, NC, and London: Duke University Press, 2006).
3. Claire Sylvia with William Novak, *A Change of Heart: A Memoir* (London: Little, Brown, 1997); Paul Pearsall, Gary E. R. Schwartz, and Linda G. S. Russek, 'Changes in Heart Transplant Recipients that Parallel the Personalities of their Donors', *Journal of Near-Death Studies*, 20 (2002), 191–206.
4. A recent article suggests there are connections between liver transplantation and personality changes in present-day China, however. Further research is needed to discover whether this is related to beliefs about the liver, emotions, and the self in the Chinese medical tradition that parallel the beliefs about the heart presented here. See http://english.people.com.cn/200602/16/eng20060216_243399.html, accessed 2 Jan. 2009.

5. "heart, n.1a", *The Oxford English Dictionary*, 2nd edn, 1989, OED Online, Oxford University Press, http://dictionary.oed.com/cgi/entry/ 50103719, accessed 3 Mar. 2008.

6. "heart, 2 and 5", *The Oxford English Dictionary*, 2nd edn, 1989, OED Online, Oxford University Press, http://dictionary.oed.com/cgi/entry/ 50103719, accessed 2 Jan. 2008.

7. See Fay Bound Alberti, 'The Emotional Heart: Mind, Body and Soul', in James Peto (ed.), *The Heart* (New Haven and London: Yale University Press, 2007), 125–42. See also Chapter 1.

8. Bound Alberti, 'The Emotional Heart'; Andrew Gregory, *Harvey's Heart: The Discovery of Blood Circulation* (Cambridge: Icon Books, 2001).

9. On the construction of the metaphor of the heart as pump, and the links between scientific metaphor and social change more generally, see Winfried Nöth (ed.), *Semiotics of the Media: State of the Art, Projects, and Perspectives* (Approaches to Semiotics, 127; Berlin: Mouton de Gruyter, 1997), 33.

10. See Otto Mayr, *Authority, Liberty and Automatic Machines in Early Modern Europe* (Baltimore: Johns Hopkins University Press, 1986); Patricia S. Churchland, *Neurophysiology: Toward a Unified Science of the Mind/Brain* (Cambridge, MA: MIT Press, 1986); William Barrett, *Death of the Soul: From Descartes to the Computer* (New York: Doubleday, 1986).

11. See Chapter 2.

12. See Chapter 1.

13. See Chapters 1 and 3.

14. See Chapter 4.

15. For example, the belief that heart complaints were 'classified as functional disorders until either the appropriate clinical–pathological relationships were established, or the requisite technology (the sphygmomanometer, the electrocardiograph) came into clinical usage' (Charles F. Wooley, *The Irritable Heart of Soldiers and the Origins of Anglo-American Cardiology: The US Civil War (1861) to World War 1 (1918)* (Aldershot: Ashgate, 2002), 67).

16. See Chapters 3 and 7.

17. On constructions of modernity, see Martin Daunton and Bernard Reiger (eds), *Meanings of Modernity: Britain from the Late Victorian Era to World War II* (Oxford: Berg, 2001); Roger Cooter, Mark Harrison, and Steve Sturdy (eds), *War, Medicine and Modernity* (Stroud: Sutton, 1998); Simon J. Williams, 'Modernity and the Emotions: Corporeal Reflections on the (Ir)rational', *Sociology*, 32 (1988), 747–69; Ian Burkitt, 'Review: Beyond the Iron Cage: Anthony Giddens on Modernity and the Self', *History of the Human Sciences*, 5 (1992), 72–9; Charles Taylor, *Sources of the Self: Making of the Modern Identity* (Cambridge: Cambridge University Press, 1992).

18. Fernando Vidal, 'Brainhood, Anthropological Figure of Modernity', *History of the Human Sciences*, 22 (2009), 5–36, at 6.

19. See Chapter 7.

20. For a recent example of the entrenchment of Cartesianism in modern ways of ordering thought and selfhood, see Paul Bloom, *Descartes' Baby: How Child Development Explains What Makes Us Human* (London: Arrow, 2005), in which 'even babies' are understood to distinguish bodies from souls. For a feminist critique of the assumptions underpinning Cartesianism, see Susan Bordo, 'Introduction', and James A. Winders, 'Writing Like a Man(?): Descartes, Science and Madness', in Susan Bordo (ed.), *Feminist Interpretations of René Descartes* (University Park, PA: Penn State Press, 1999),1–29; 114–41. On 'Emotionology', see Peter N. Stearns and Carol Z. Stearns, 'Emotionology: Clarifying the History of Emotions and Emotional Standards', *American Historical Review*, 90 (1985), 813–36. The extent to which emotions can be said to exist outside their performances is another matter. See Fay Bound, 'Emotion in Early Modern England: Performativity and Practice at the Church Courts of York, *c.*1660–1760' (DPhil, York, 2000), introduction.

21. On the rise of experimental physiology and emotion research in the laboratory, see Otniel E. Dror, 'The Affect of Experiment: The Turn to Emotions in Anglo-American Physiology, 1900–1940', *Isis*, 90 (1999), 205–37.

22. For an introduction, see Antonio Damasio, *The Feeling of What Happens: Body, Emotion and the Making of Consciousness* (New York: Harcourt, 2000).

23. The definition of 'death' remains culturally and medically contentious. See T. T. Randell, 'Medical and Legal Considerations of Brain Death', *Acta Anaesthesiologica Scandinavica*, 48 (2004), 139–44.

24. Charles G. Gross, 'Aristotle on the Brain', *Neuroscientist*, 1 (1995), 245–50. See also Chapter 7.

25. See Chapter 5.

26. Christopher Lawrence and George Weisz (eds), *Greater than the Parts: Holism in Biomedicine, 1920–1950* (New York and Oxford: Oxford University Press, 1998).

27. On Romanticism and emotion, see Andrew M. Stauffer, *Anger, Revolution and Romanticism* (Cambridge Studies in Romanticism, 62; Cambridge: Cambridge University Press, 2005); David Vallins, *Coleridge and the Psychology of Romanticism: Feeling and Thought* (Basingstoke: Macmillan, 2000); Ole Martin Høystad, *A History of the Heart* (London: Reaktion, 2007); Kirstie Blair, *Victorian Poetry and the Language of the Heart* (Oxford and New York: Oxford University Press, 2006).

28. The best-known exponents of this perspective are the physician Andrew Armour and the Institute of Heartmath: http://www.heartmath.org, accessed 2 Jan. 2009. See also Doc Lew Childre and

Howard Martin with Donna Beech, *The HeartMath Solution: Proven Techniques for Developing Emotional Intelligence* (London: Piatkus, 1999).

29. Joseph S. Alter, *Asian Medicine and Globalization* (Philadelphia: University of Pennsylvania Press, 2005).

30. On the future directions of mind-brain work, see R. L. Solso (ed.), *Mind and Brain Sciences in the 21st Century* (Cambridge, MA: MIT Press, 1997).

31. For an introduction to some of the main ethical issues, see Arthur L. Caplan, *Am I My Brother's Keeper? The Ethical Frontiers of Biomedicine* (Bloomington, IN: Indiana University Press, 1997).

32. Recent work by cultural historians includes Gail Kern Paster et al. (eds), *Reading the Early Modern Passions: Essays in the Cultural History of Emotion* (Philadelphia: University of Philadelphia Press, 2004). For a review of the field in relation to medical and cultural history, see Fay Bound Alberti, 'Introduction: Emotion Theory and Medical History', in Fay Bound Alberti (ed.), *Medicine, Emotion and Disease 1700–1950* (New York and Basingstoke: Palgrave Macmillan, 2006), xiii–xxviii.

33. Lucien Febvre, 'La Sensibilité et l'histoire: Comment reconstituer la vie affective d'autrefois?', *Annales d'histoire sociale*, 3 (1941), 5–20, produced in English as 'Sensibility and History: How to Reconstitute the Emotional Life of the Past', in Peter Burke (ed.), *A New Kind of History: From the Writings of Febvre*, trans. K. Folca (London: Routledge and Kegan Paul, 1973), 12–26; Stearns and Stearns, 'Emotionology'; eid., *Anger: The Struggle for Emotional Control in America's History* (Chicago and London: Chicago University Press, 1986); Peter N. Stearns, *Jealousy: The Evolution of an Emotion in American History* (New York and London: New York University Press, 1989); id., *Battleground of Desire: The Struggle for Self-Control in Modern America* (New York and London: New York University Press, 1999); Bound Alberti (ed.), *Medicine, Emotion and Disease*.

34. Stearns, *Jealousy*; Stearns and Stearns, *Anger*; Joanna Bourke, *Fear: A Cultural History* (London: Virago Press, 2005).

35. For an interesting recent article on collective sentiment and emotional sharing, see Bernard Rimé, 'The Social Sharing of Emotion as an Interface between Individual and Collective Processes in the Construction of Emotional Climates', *Journal of Social Issues*, 63 (2007), 307–22.

36. Stearns and Stearns, 'Emotionology', and Barbara H. Rosenwein, 'Worrying about Emotions in History', *American Historical Review*, 107 (2002), 821–45, and ead., *Emotional Communities in the Early Middle Ages* (Ithaca, NY: Cornell, 2007); William Reddy, *Navigation of Feeling: A Framework for the History of Emotions* (Cambridge: Cambridge University Press, 2001), 5.

37. Important examples from anthropology include Jean Briggs, *Never in Anger* (Cambridge, MA: Harvard University Press; London: Oxford

University Press, 1970); Robert I. Levy, *Tahitians: Mind and Experience in the Society Islands* (Chicago and London: University of Chicago Press, 1973); Richard A. Shweder and R. A. LeVine (eds), *Culture Theory: Essays on Mind, Self and Emotion* (Cambridge: Cambridge University Press, 1984). On the psychology of emotions, see Jochen Musch and Karl Christoph Klauer (eds), *The Psychology of Evaluation: Affective Processes in Cognition and Emotion* (Mahweh, NJ, and London: Erlbaum, 2003); Robert Plutchik, *Emotions and Life: Perspectives from Psychology, Biology and Evolution* (Washington: American Psychological Association, 2003). On the sociology of emotion, see David D. Franks and E. Doyle McCarthy (eds), *The Sociology of Emotions: Original Essays and Research Papers* (Contemporary Studies in Sociology, 9; Greenwich, CT, and London: JAI, 1989); Jack Barbalet (ed.), *Emotions and Sociology* (Sociological Review Monograph Series; Oxford: Blackwell, 2002); Michael Lewis and Jeanette M. Haviland-Jones (eds), *Handbook of Emotions* (New York: Guildford Press, 2000).

38. See Otniel E. Dror, 'The Scientific Image of Emotion: Experience and Technologies of Inscription', *Configurations*, 7 (1999), 355–401.

39. See the discussion in Chapters 3 and 7.

40. On the challenges of comparative chronological studies, see Bound, 'Emotion in Early Modern England', 5–6, and Bound Alberti, 'Introduction'.

41. Herman A. Snellen, *History of Cardiology: A Brief Outline of the 350 Years' Prelude to an Explosive Growth* (Rotterdam: Donker, 1984); Louis J. Acierno, *History of Cardiology* (Parthenon, 1993); Bruce W. Fye, *American Cardiology: The History of a Specialty and its College* (Baltimore and London: Johns Hopkins University Press, 1996); Peter Robert Fleming, *A Short History of Cardiology* (Amsterdam: Rodopi, 1997); William F. Bynum, Christopher Lawrence, and Vivian Nutton (eds), *The Emergence of Modern Cardiology* (Medical History Supplement No. 5; London: Wellcome Institute, 1995); Wooley, *Irritable Heart*.

42. Harris B. Shumacker, *The Evolution of Cardiac Surgery* (Bloomington, IN: Indiana University Press, 1992); Karel B. Absolon and Mohammad A. Naficy, *First Successful Cardiac Operation in a Human, 1896: A Documentation: The Life, the Times, and the Work of Ludwig Rehn (1849–1930)* (Rockville, MD: Kabel, 2002); Robert Galloway Richardson, *Heart and Scalpel: A History of Cardiac Surgery* (London: Quiller Press, 2001); Peter Hawthorne, *The Transplanted Heart: The Incredible Story of the Epic Heart Transplant Operations by Professor Christiaan Barnard and his Team* (Johannesburg: Hugh Keartland, 1968); Nicholas L. Tilney, *Transplant: From Myth to Reality* (New Haven: Yale University Press, 2003).

43. Louisa Young, *The Book of the Heart* (London: Flamingo, 2002); Milad Doueihi, *A Perverse History of the Human Heart* (Cambridge, MA, and London: Harvard University Press, 1997); Noubar Boyadjian, *The*

Heart: Its History, its Symbolism, its Iconography and its Diseases (Antwerp: Esco, 1980); Robert A. Erickson, *The Language of the Heart, 1600–1750* (Philadelphia: University of Pennsylvania Press, 1997); Blair, *Victorian Poetry*, reviewed by Fay Bound Alberti in *Medical History*, 52 (2008), 156–7. See also Peto (ed.), *The Heart*.

44. Anna Wierzbicka, *Emotions across Languages and Cultures: Diversity and Universals* (Cambridge: Cambridge University Press, 1999).

45. On the use of emotional rhetoric in the courtroom, in marriage disputes, and in love letters, see Fay Bound, 'An "Angry and Malicious Mind"? Narratives of Slander at the Church Courts of York, *c.*1660–*c.*1760', *History Workshop Journal*, 56 (2003), 59–77; ead., 'Writing the Self? Love and the Letter in England, *c.*1660–*c.*1760', *Literature and History*, 11 (2002), 1–19; ead., 'An "Uncivill" Culture: Marital Violence and Domestic Politics in York, *c.*1660–*c.*1760', in Mark Hallett and Jane Rendall (eds), *Eighteenth-Century York: Culture, Space and Society* (York: Borthwick, 2003), 50–8. For an introduction to narratives of illness and suffering, see the discussion in Chapter 6.

46. Thomas Dixon, *From Passions to Emotions: The Creation of a Secular Psychological Category* (Cambridge: Cambridge University Press, 2003), introduction.

47. Bound, 'Emotion in Early Modern England', 24–7; Dixon, *Passions to Emotions*, introduction.

48. For a development of this theme in relation to medical history in particular, see Bound Alberti, 'Introduction'.

49. Most psychological studies suggest there are five basic emotions—happiness, anxiety, sadness, anger, and disgust—all of which are believed to have a survival value, and to be linked to evolutionary purposes. Others, including Paul Ekman, argue that there are six basic emotions, the surprise addition being 'surprise'. For an introduction, see Paul Ekman, *The Nature of Emotion: Fundamental Questions* (New York: Oxford University Press, 1995).

50. See the discussion in Chapter 3.

51. Terence Irwin, *The Development of Ethics: A Historical and Critical Study*, i. *From Socrates to the Reformation* (Oxford and New York: Oxford University Press, 2007); Susan James, *Passion and Action: The Emotions in Seventeenth-Century Philosophy* (Oxford: Oxford University Press, 1997), 96–7; Dixon, *Passions to Emotions*.

52. See Chapter 6 and Conclusion.

53. See Blair, *Victorian Poetry,* introduction. See also Chapter 6.

54. See Chapters 3 and 4.

55. See Chandak Sengoopta, *The Most Secret Quintessence of Life: Sex, Glands, and Hormones, 1850–1950* (Chicago and London: University of Chicago Press, 2006).

56. The 23-year-old heart transplant patient's first reaction on seeing the old heart was disgust, but later she described the experience as slightly surreal. She said: 'Because it was mine, I was like, wow, that's my heart, I just couldn't stop grinning. It's odd to think that I was stood here alive and that was part of me once upon a time.' See: http://entertainment.timesonline.co.uk/tol/arts_and_entertainment/ visual_arts/article2388302.ece; accessed 23 Jan. 2008.

57. See Bound Alberti, 'The Emotional Heart'.

CHAPTER 1

1. 'heart n.' *The Concise Oxford English Dictionary*, 12th edn, ed. Catherine Soanes and Angus Stevenson (Oxford: Oxford University Press, 2008). Oxford Reference Online. Oxford University Press, http://www. oxfordreference.com/views/ENTRY.html?subview= Main&entry=t23.e25521, accessed 17 Feb. 2009. Sections of the argument presented here have been rehearsed in Fay Bound Alberti, 'The Emotional Heart: Mind, Body and Soul', in James Peto (ed.), *The Heart* (New Haven and London: Yale University Press, 2007), 125–42.

2. Answers vary, but commonly suggest approx. 12,000 pints. For an oft-cited website example of such mathematical concerns, see http://www. smm.org/heart/heart/top.html, accessed 2 Jan. 2009.

3. One example of a successful textbook on heart disease in the West is Desmond Gareth Julian (ed.), *Diseases of the Heart* (London: Saunders, 1996).

4. See Noubar Boyadjian, *The Heart: Its History, its Symbolism, its Iconography and its Diseases* (Antwerp: Esco, 1980); Robert A. Erickson, *The Language of the Heart, 1600–1750* (Philadelphia: University of Pennsylvania Press, 1997); and Emily Jo Sargent, 'The Sacred Heart', in Peto (ed.), *The Heart*, 102–14.

5. Examples include Joseph Rhawn's expansive *Neuropsychiatry, Neuropsychology, and Clinical Neuroscience: Emotion, Evolution, Cognition, Language, Memory, Brain Damage, and Abnormal Behaviour*, 2nd edn (Baltimore: Williams and Wilkins, 1996); Allan N. Schore, *Affect Regulation and the Origin of the Self: The Neurobiology of Emotional Development* (Hillsdale, NJ, and Hove: L. Erlbaum Associates, 1994).

6. See Chapter 7.

7. Ian Burkitt, *Bodies of Thought: Embodiment, Identity and Modernity* (London: Sage, 1999); id., *Social Selves: Theories of Self and Society*, 2nd edn (Los Angeles and London: Sage, 2008).

8. For an introduction to these themes, see Thomas Dixon, *From Passions to Emotions: The Creation of a Secular Psychological Category* (Cambridge: Cambridge University Press, 2003).

9. See Fay Bound Alberti, 'Introduction: Medical Theory and Emotion History', in ead. (ed.), *Medicine, Emotion and Disease, 1700–1950* (Basingstoke: Palgrave, 2006), xiii–xxviii.

10. See Galen, *On the Passions and Errors of the Soul*, trans. Paul W. Harkins ([Columbus]: Ohio State University Press, 1963), and Oswei Temkin, *Galenism: Rise and Decline of a Medical Philosophy* (Ithaca, NY, and London: Cornell University Press, 1973), ch. 1.

11. See Helkiah Crooke, MD, *Mikrokosmographia: A Description of the Body of Man* (London, 1615), and the discussion in John Bernard Bamborough, *The Little World of Man* (London and New York: Longmans, Green, 1952), ch. 1.

12. For a dated, but still relevant, synopsis, see Lily Campbell, *Shakespeare's Tragic Heroes: Slaves of Passion* (London: Methuen, 1930), 51.

13. Temkin, *Galenism*, ch. 1.

14. Thomas Hobbes, *Leviathan*, ed. Richard Tuck (1651; repr. Cambridge: Cambridge University Press, 1991), 53.

15. Thomas Wright, *Passions of the Minde in Generall* (1601, 1604; repr. Urbana, IL: University of Illinois Press, 1971), 64.

16. Wright, *Passions of the Minde*, 40.

17. Lelland Joseph Rather, 'Old and New Views of the Emotions and Bodily Changes: Wright and Harvey versus Descartes, James and Cannon', *Clio Medica*, 1 (1965), 1–25, at 4.

18. Robert Burton, *Anatomy of Melancholy* (1621; repr. New York: New York Review of Books, 2001), pt 1. See also Akhito Suzuki, 'Mind and its Disease in Enlightenment British Medicine' (PhD, University College London, 1992), 53.

19. Bamborough, *Little World of Man*, 64.

20. Wright, *Passions of the Minde*, 71–3.

21. Levinus Lemnius, *The Secret Miracles of Nature* (London, 1658), 274, cited in Bamborough, *Little World of Man*, 64.

22. Burton, *Anatomy of Melancholy*, 152–3.

23. Pierre de la Primaudaye, *The French Academie*, trans. Thomas Bowes et al. (London: Printed for T. Adams, 1618 [1577]), 471.

24. John Downame, *A Treatise of Anger* (London: Printed by T. E. for William Welby, 1609), 3.

25. Eric Jager, *The Book of the Heart* (Chicago and London: University of Chicago Press, 2000), p. xv.

26. Wright, *Passions of the Minde*, 11, 8.

27. See the discussion in Rather, 'Old and New', 4.

28. Walter Charleton, *Natural History of the Passions* ([London] In the Savoy: Printed by T. N. for J. Magnes, 1674), 70.

29. Downame, *Treatise of Anger*, 56.

30. Charleton, *Natural History*, 151. On the 'animal spirits', see L. Stephen Jacyna, 'Animal Spirits and Eighteenth-Century British Medicine', in

Yosio Kawakita, Shizu Sakai, and Yasuo Otsuka (eds), *The Comparison between Concepts of Life-Breath in East and West*, (Tokyo and St Louis: Ishiyaku EuroAmerica, 1995).

31. Charleton, *Natural History*, 151.

32. William Harvey, *Exercitatio Anatomica de Motu Cordis et Sanguinus in Animalibus* (Frankfurt: Sumptibus Guilielmi Fitzeri 1628); Gregory, *Harvey's Heart*.

33. Isaac Newton, *Philosophiae Naturalis Principia Mathematica* (London: J. Streater for Royal Society, 1687); Niccolò Guicciardini, *Reading the* Principia: *The Debate on Newton's Mathematical Methods for Natural Philosophy from 1687 to 1736* (Cambridge: Cambridge University Press, 1999).

34. See George S. Rousseau, *Enlightenment Borders: Pre- and Post-Modern Discourses: Medical and Scientific* (Manchester: Manchester University Press, 1991), 125.

35. Rousseau, *Enlightenment Borders*, 84.

36. On the difficulties of making clean-cut distinctions between practitioners in each of these fields, see Rousseau, *Enlightenment Borders*, ch. 5, esp. p. 120.

37. Thomas Willis, *Dr Willis's Practice of Physick* (London: T. Dring, C. Harper and J. Leigh, 1684); Archibald Pitcairne, *Dissertationes Medicae* (Edinburgh: Robert Freebairn, 1713); Friedrich Hoffman, *A System of the Practice of Medicine; From the Latin of Dr Hoffman* (2 vols; London: J. Murray and J. Johnson, 1783); Albrecht Von Haller, *First Lines of Physiology*, ed. Lester S. King (2 vols; 1786; facs. repr. New York: Johnson Reprint Corp, 1966). Von Haller's work underwent significant and posthumous alterations after its first publication in 1747. Some of these alterations are acknowledged briefly in this chapter through comparative use of the 1786 edition, as reprinted by Johnson Reprint Corp. in 1966; the 1779 edition, *First Lines of Physiology, by the Celebrated Baron Albertus Haller; Translated from the Correct Latin Edition, Printed under the Inspection of William Cullen* (Edinburgh: Charles Elliot, 1779); and the 1803 first American edition of *First Lines of Physiology, to which Is Added a Translation of the Index, Composed for the Edinburgh Edition, Printed under the Inspection of Dr William Cullen* (Troy: Obadiah Penniman & Co.; Albany: C. R. & G. Webster, 1803). See also William Cullen, *Synopsis Nosologiae Methodicae* (Edinburgh, 1769), Electronic resource: *The Eighteenth Century*, reel 6678, no. 05; I. A. Bowman, 'William Cullen (1710–1790) and the Primacy of the Nervous System' (PhD, University of Indiana, 1975), 12.

38. See Bowman, 'William Cullen (1710–1790)', 12.

39. René Descartes, *Les Passions de l'âme* (Amsterdam: [Henry Le Gras] chez Louys Elzevier, 1649). See Richard Olson, *The Emergence of the Social Sciences 1642–1792* (New York: Twayne Publishers; Oxford: Macmillan,

1993), 39; Roy Porter, *Enlightenment: Britain and the Creation of the Modern World* (London: Allen Lane, 2000), 139.

40. See Stanley W. Jackson, *Melancholia and Depression: From Hippocratic Times to Modern Times* (New Haven and London: Yale University Press, 1986), 21.

41. See Rather, 'Old and New', 1.

42. Susan James, *Passion and Action: The Emotions in Seventeenth-Century Philosophy* (Oxford: Oxford University Press, 1997), 96–7.

43. James, *Passion and Action*, 97.

44. Rather, 'Old and New', 12.

45. See Elizabeth L. Haigh, 'Vitalism, the Soul and Sensibility: The Physiology of Théophile Bordeu', *Journal of the History of Medicine and Allied Sciences*, 1 (1976), 30–41.

46. Haigh, 'Vitalism', 30.

47. See Johanna Geyer-Kordesch, 'Georg Ernst Stahl's Radical Pietist Medicine and its Influence on the German Enlightenment', in Andrew Cunningham and Roger French (eds), *The Medical Enlightenment of the Eighteenth Century* (Cambridge and New York: Cambridge University Press, 1990), 68.

48. Geyer-Kordesch, 'Georg Ernst Stahl's Radical Pietist Medicine'.

49. Haigh, 'Vitalism', 31.

50. Harvey, *De Motu Cordis*; John Locke, *An Essay Concerning Human Understanding* (1697), ed. A. S. Pringle-Pattison, introduction by Diane Collinson (Wordsworth Classics of World Literature; Ware: Wordworth Editions, 1998). Much has been written about the domination of nerve theory in eighteenth-century diagnoses. For reprints of, and reflections on, his now classic essays on the subject, see George S. Rousseau, *Nervous Acts: Essays on Literature, Culture and Sensibility* (Basingstoke and New York: Palgrave Macmillan, 2004).

51. Porter, *Enlightenment*, 139. See also Roy Porter (ed.), *George Cheyne: The English malady* (1733; facs. repr. London: Tavistock/Routledge, 1991). On Cheyne's struggle with his weight and attempts to live rationally, see Rousseau, *Enlightenment Borders*, ch. 4.

52. Pitcairne, *Dissertationes*, discussed in Jackson, *Melancholia and Depression*, 117. See also Anita Guerrini, 'Isaac Newton, George Cheyne and the "Principia Medicinae"', in Roger Kenneth French and Andrew Wear (eds), *The Medical Revolution of the Seventeenth Century* (Cambridge: Cambridge University Press, 1989), 222–45, at 224.

53. Jackson, *Melancholia and Depression*, 288.

54. Friedrich Hoffman, *Fundamenta Medicinae* (1695), trans. Lester S. King (London: Macdonald & Co., 1971), 12. See also Jackson, *Melancholia and Depression*, 118.

55. For an introduction to Boerhaave's work, see Andrew Cunningham, 'Medicine to Calm the Mind: Boerhaave's Medical System, and Why

it was Adopted in Edinburgh', in Cunningham and French (eds), *The Medical Enlightenment*, 40.

56. Cunningham, 'Medicine to Calm the Mind', 43, 65.
57. Jackson, *Melancholia and Depression*, 119.
58. Jackson, *Melancholia and Depression*, 120.
59. Von Haller, *First Lines* (1803 edn), 9.
60. Von Haller, *First Lines* (1801 edn), 201.
61. For an introduction see Rousseau, *Nervous Acts*, introduction.
62. William Clark, MD, *A Medical Dissertation Concerning the Effects of the Passions on Human Bodies* (Bath and London: for W. Frederick, 1752), 38–40.
63. Clark, *Medical Dissertation*, 40.
64. An animist approach, as seen in the work of Robert Whytt, maintained that emotions increased bodily sensitivity, and so impacted on the stimulation produced in the heart by the blood. For an insightful interpretation, see Roger Kenneth French, *The History of the Heart: Thoracic Physiology from Ancient to Modern Times* (Aberdeen: Equipress, 1979), 74.
65. Edwin Clarke and L. Stephen Jacyna, *Nineteenth-Century Origins of Neuroscientific Concepts* (Berkeley and Los Angeles, and London: University of California Press, 1987), 4.
66. Robert Whytt, *Observations on the Nature, Causes and Cure of those Disorders Which Have Been Commonly Called Nervous, Hypochondriac or Hysteric* (London: T. Maiden, 1797), 42.
67. Von Haller, *First Lines* (1803 edn), 201.
68. Von Haller, *First Lines* (1803 edn), 214.
69. Von Haller, *First Lines* (1803 edn), 271–2.
70. Von Haller, *First Lines* (1803 edn), 275.
71. See Beate Gundert, 'Soma and Psyche in Hippocratic Medicine', in John P. Wright and Paul Potter (eds.), *Psyche and Soma: Physicians and Metaphysicians on the Mind–Body Problem from Antiquity to Enlightenment* (Oxford: Oxford University Press, 2000), 28.
72. Von Haller, *First Lines* (1803 edn), 276.
73. Dr Corp, MD of Bath, *An Essay on the Changes Produced in the Body by the Operations of the Mind* (London: James Ridgway, 1791).
74. Thomas Cogan, MD, *A Treatise on the Passions and Affections of the Mind* (5 vols; London: T. Cadell and W. Davies, 1813), ii, pt 1: 'On the Pursuit of Well-Being', 6.
75. Von Haller, *First Lines* (1786 edn), 85. In later editions, *First Lines* placed more emphasis on the nature of individual 'temperament' (discussed in more detail at n. 76) and the movements of the fluids around the body. See *First Lines* (1803 edn), 82.
76. On 'temperament', see George S. Rousseau, 'Temperament and the Long Shadow of Nerves in the Eighteenth Century', in Harry Whitaker,

Christopher Upham Murray Smith, and Stanley Finger (eds), *Brain, Mind and Medicine: Essays in Eighteenth-Century Neuroscience* (New York: Springer, 2007).

77. See Chapter 2.

78. John Bond, MD, *An Essay on the Incubus or Night-Mare* (London: D. Wilson & T. Durham, 1753), 5.

79. William Buchan, *Domestic Medicine: or, a Treatise on the Prevention and Cure of Diseases by Regimen and Simple Medicines* (1772 edn; facs. repr. New York and London: Garland, 1985), 168.

80. Buchan, *Domestic Medicine*, 168; William Rowley, *A Treatise on Female, Nervous, Hysterical, Hypochondriacal, Bilious, Convulsive Diseases; Apoplexy and Palsy, with Thoughts on Madness, Suicide, &c* (London, 1788), 1.

81. Bond, *Essay on the Incubus*, 46.

82. Bond, *Essay on the Incubus*, 50.

83. Marco Piccolinoa, 'Luigi Galvani and Animal Electricity: Two Centuries after the Foundation of Electrophysiology', *Trends in Neurosciences*, 20 (1997), 443–8.

84. Alison Winter, *Mesmerized: Powers of Mind in Victorian Britain* (Chicago and London: University of Chicago Press, 1998), 2–33. The writer, economist, and philosopher Harriet Martineau famously used mesmerism to combat her ailments, as discussed in Chapter 6.

85. Winter, *Mesmerized*, 37.

86. Winter, *Mesmerized*, 28.

87. See Wright and Potter (eds), *Psyche and Soma*, and George S. Rousseau and Roy Porter, 'Introduction: Toward a Natural History of Mind and Body', in George S. Rousseau (ed.), *The Languages of Psyche: Mind and Body in Enlightenment Thought* (Berkeley and Los Angeles, and Oxford: University of California Press, 1990), 3–44. For more recent discussion, see Paul S. Macdonald, *History of the Concept of Mind: Speculations about Soul, Mind and Spirit from Homer to Hume* (Aldershot: Ashgate, 2007).

88. See, e.g. the Paracelsian physician Jean-Baptiste Van Helmont, *The Spirit of Diseases; or, Diseases from the Spirit... Wherein is Shewed How Much the Mind Influenceth the Body in Causing and Curing of Diseases* (London: Sarah Howkins, 1694), available as an electronic resource at Early English Books Online: http://eebo.chadwyck.com.

89. Van Helmont, *The Spirit of Diseases*, 31.

90. Van Helmont, *The Spirit of Diseases*, 47.

91. Van Helmont, *The Spirit of Diseases*, 86–7.

92. Rousseau and Porter, 'Introduction: Toward a Natural History of Mind and Body', 4, n. 2.

93. On hypochondria, see Frederik Albritton Johnson, 'Physiology of Hypochondria in Eighteenth-Century Britain', in Christopher E. Forth and Ana Carden-Coyne (eds), *Cultures of the Abdomen: Diet, Digestion*

and Fat in the Modern World (New York and Basingstoke: Palgrave, 2005); Vladan Starcevic and Don R. Lipsitt (eds), *Hypochondriasis: Modern Perspectives on an Ancient Malady* (Oxford: Oxford University Press, 2001); Jeremy Schmidt, *Melancholy and the Care of the Soul: Religion, Moral Philosophy and Madness in Early Modern England* (Aldershot and Burlington, VT: Ashgate, 2007).

94. Marie François Xavier Bichat, *General Anatomy, Applied to Physiology and the Practice of Medicine*, rev. George Calvert (London: Printed for the Translator, 1824); Edward Shorter, *From Paralysis to Fatigue: A History of Psychosomatic Illness in the Modern Era* (New York: Free Press, 1992), 18-19.

95. See Friedrich Hoffman, *A System of the Practice of Medicine: From the Latin of Dr Hoffman* (2 vols; London: J. Murray and J. Johnson, 1783). For a discussion of Hoffman's theory in contemporary context, see Andrew Wear, *Medicine in Society: Historical Essays* (Cambridge: Cambridge University Press, 1992), 163.

96. William Cullen, *Synopsis Nosologiae Methodicae* (Edinburgh: 1769), electronic resource: *The Eighteenth Century*, reel 6678, no. 05; Shorter, *Paralysis to Fatigue*, 19.

97. Michael Neve, 'Neurosis', *Lancet*, 363/9415 (2004), 1170; Eduard Hitschmann, *Freud's Theories of the Neuroses ... Authorized Translation by Dr C. R. Payne etc [With a Bibliography]* (New York: Moffat, Yard and Co., 1917).

98. George Miller Beard, *Sexual Neurasthenia (Nervous Exhaustion): Its Hygiene, Causes, Symptoms and Treatment with a Chapter on Diet for the Nervous*, ed. A. D. Rockwell, 5th edn (New York: E. B. Treat, 1898). See Doris Kaufmann, 'Neurasthenia in Wilhelmine Germany: Culture, Sexuality and the Demands of Nature', in Marijke Gikswijt-Hofstra and Roy Porter (eds), *Cultures of Neurasthenia from Beard to the First World War* (Wellcome Institute Series in the History of Medicine, 63; Amsterdam: Rodopi, 2001), 161-76.

99. Chandak Sengoopta, 'A Mob of Incoherent Symptoms: Neurasthenia in British Medical Discourse, 1860-1920', in Gijswijt-Hofstra and Porter (eds.), *Cultures of Neurasthenia*, 97-115, at 101.

100. Lelland Joseph Rather, *Mind and Body in Eighteenth-Century Medicine: A Study Based on Jerome Gaub's* De Regimine Mentis (London: Wellcome Historical Medical Library, 1965).

101. John Petvin, *Letters Concerning Mind: To Which is Added, A Sketch of Universal Arithmetic; Comprehending the Differential Calculus, and the Doctrine of Fluxions* (London: John and James Rivington, 1750); John Richardson, *Thoughts upon Thinking; or, A New Theory of the Human Mind; Wherein a Physical Rationale of the Formation of our Ideas, the Passions, Dreaming and Every Faculty of the Soul is Attempted upon Principles Entirely New* (London: J. Dodsley, 1755); John Rotherham, *On the*

Distinction between the Soul and the Body (London: J. Robson, 1760). See the discussion in Rousseau and Porter, 'Introduction: Toward a Natural History of Mind and Body', 4, n. 2.

102. See Chapter 7.

103. For an introduction, see Brendan Sweetman (ed.), *The Failure of Modernism: The Cartesian Legacy and Contemporary Pluralism* (Mishawaka, IN: American Maritain Association, 1999).

104. See Chapter 6.

105. See, e.g., William Nisbet Chambers, 'Emotional Stress in the Precipitation of Congestive Heart Failure', *Psychosomatic Medicine*, 15 (1953), 38–60. On 'romantic science', see Martin Halliwell, *Romantic Science and the Experience of Self: Transatlantic Crosscurrents from William James to Oliver Sacks* (Aldershot and Brookfield, VT: Ashgate, 1999); Clarke and Jacyna, *Nineteenth-Century Origins*, 1–4.

106. Rhodri Hayward, *Self-Cures: Psychology and Medicine in Modern Britain* (forthcoming), introduction. Thanks to Rhodri for allowing me to quote from his work prior to its publication.

107. Edward Stainbrook, 'Psychosomatic Medicine in the Nineteenth Century', *Psychosomatic Medicine*, 14 (1952), 211–27, at 211.

108. Daniel Hack Tuke, *Illustrations of the Influence of the Mind upon the Body in Health and Disease: Designed to Elucidate the Action of the Imagination* (London: J. and A. Churchill, New Burlington Street, 1872; repr. Philadelphia: H. C. Lea's Son and Co., 1884), 5.

109. See Julius Rocca, 'William Cullen (1710–1790) and Robert Whytt (1714–1766) on the Nervous System', in Whitaker, Smith, and Finger (eds.), *Brain, Mind and Medicine*, 85–98.

110. Arthur F. Hughes, 'A History of Endocrinology', *Journal of the History of Medicine and Allied Sciences*, 32 (1977), 292–313; Diane Hall, 'Biology, Sex Hormones, and Sexism in the 1920s', *Philosophical Forum*, 5 (1974), 81–96; Michael Bliss, *The Discovery of Insulin* (Chicago: University Chicago Press, 1982); Chandak Sengoopta, *The Most Secret Quintessence of Life: Sex, Gland and Hormones, 1850–1950* (Chicago and London: University of Chicago Press, 2006), introduction.

111. Ernst Starling, 'Croonian Lecture: On the Chemical Correlation of the Functions of the Body', *Lancet*, 2 (1905), 339–41, at 339.

112. W. W. Meissner, 'Psychoanalysis and the Mind-Body Relation: Psychosomatic Perspectives', *Bulletin of the Menninger Clinic*, 70 (2006), 295–315, and Brian Dolan, 'Soul Searching: A Brief History of the Mind-Body Debate in the Neurosciences', *Neurosurgical Focus*, 23 (2007), 1–7.

113. See http://www.heartmath.org/research/our-heart-brain.html, accessed 3 Mar. 2008.

114. http://www.heartmath.org/research/science-of-the-heart/index.html, accessed 3 Mar. 2008.

115. Pratibha Mamgain, 'A Critical Study on the Concept of Ischaemic
Heart Disease in Ayurveda', *Ancient Science of Life*, 13 (1993–4), 102–10;
H. S. Wasir, *Traditional Wisdom for Heart Care* (New Delhi: Vikas, 1995);
Skya Abbate, *Advanced Techniques in Oriental Medicine* (Stuttgart and New
York: George Thieme Verlag, 2006).

116. John H. K. Vogel and Mitchell W. Krucoff (eds), *Integrative Cardiology: Complementary and Alternative Medicine for the Heart* (New York:
McGraw-Hill Medical; London: McGraw Hill [distributor], 2007).

117. In 2008, for instance, the British Medical Holistic Association held its
annual conference on the theme of 'the relational heart', exploring
the ways lifestyle and heart health could intersect with therapeutic practices. See http://www.bhma.org/new_site/past_conferences/
Relational_Heart2008Conf.pdf, accessed 2 Jan. 2008.

CHAPTER 2

1. *The Philosophy of Medicine or, Medical Extracts on the Nature of Health
and Disease, Including the Laws of the Animal Oeconomy and the Doctrines
of Pneumatic Medicine* (London: C. Whittingham for T. Cox etc.,
1799–1800), 402.

2. *The Works of John Hunter, F.R.S.*, ed. James F. Palmer (4 vols; London: Longman, Rees, Orme, Browne, Green and Longman, 1835),
i. 131. Palmer was a senior surgeon to St George's and St James' Dis-
pensary, and Fellow of the Royal Medical and Chirurgical Society of
London.

3. Autopsy report reproduced by Brian Livesley in 'The Spasms of John
Hunter: A New Interpretation', *Medical History*, 17 (1973), 70–5, at 70.

4. Everard Home, cited in Livesley, 'Spasms', 70.

5. Everard Home, cited in Livesley, 'Spasms', 70.

6. On Hunter's museum collection, see Fay Bound Alberti and Samuel
J. M. M. Alberti, 'Review: The Hunterian Museum at the Royal
College of Surgeons of England, London', *Bulletin of the History of
Medicine*, 80/3 (2006), 571–3.

7. Wendy Moore, *The Knife Man: The Extraordinary Life and Times of John
Hunter, Father of Modern Surgery* (London: Bantam, 2005).

8. See L. Stephen Jacyna, 'Images of John Hunter in the Nineteenth
Century', *History of Science*, 21 (1983), 85–108, at 87.

9. See *Works of John Hunter*, ed. Palmer, 52–3.

10. Moore, *The Knife Man*. See, e.g. pp. 45, 125, 346, 347.

11. Dr V. Knox, cited in Thomas J. Pettigrew, 'John Hunter: From the
Medical Portrait Gallery', *Lancet*, 815 (1939), 119–20, at 119.

12. See Chapter 1.

13. William L. Proudfit, 'John Hunter: On Heart Disease', *British Heart
Journal*, 56 (1986), 109–14, at 112; J. Fothergill, 'Farther Account of

the Angina Pectoris', *Medical Observations and Inquiries by a Society of Physicians in London*, 5 (1776), 252–8.

14. For a recent account of the breadth of Hunter's scientific preparations, particularly in relation to the history of the nerves, see James L. Stone, James T. Goodrich, and George R. Cybulski, 'John Hunter's Contributions to Neuroscience', in Harry Whitaker, C. U. M. Smith, and Stanley Finger (eds), *Brain, Mind and Medicine: Essays in Eighteenth-Century Neuroscience* (Springer: Birmingham, 2007), 67–84.

15. See Chapter 3.

16. For a history of 'angina' and its terminology, see John Forbes, Alexander Tweedie, and John Conolly (eds), *The Cyclopaedia of Practical Medicine* (London: Sherwood, Gilbert and Piper, 1833), 81. On William Heberden, see 'Some Account of a Disorder of the Breast', *Medical Transactions of the College of Physicians,* 2 (1772), 59–67; repr. in his *Commentaries on the History and Cure of Diseases*, 2nd edn (London: T. Payne, 1803). I am grateful to Sir Christopher Booth for discussions on this subject.

17. Other heart diseases to be classified included pericarditis (1799) and endocarditis (1809). See Joshua O. Leibowitz, *The History of Coronary Heart Disease* (London: Wellcome Institute, 1970). Further examples of this genre include Frederick A. Willius and Thomas Keys (eds), *Cardiac Classics* (London: Henry Kipton, 1941); Terence East, *The Story of Heart Disease* (London: William Dawson, 1958); P. R. Fleming, *A Short History of Cardiology* (Amsterdam: Rodopi, 1997).

18. Heberden, *Commentaries*, 59.

19. Heberden, *Commentaries*, 81; Charles F. Wooley, *The Irritable Heart of Soldiers and the Origins of Anglo-American Cardiology: The US Civil War (1861) to World War I (1918)* (Aldershot: Ashgate, 2002), 75. On the similar etymological origins of 'anxiety' to describe a pressing sensation within the chest, see Fay Bound, 'Anxiety', *Lancet*, 363/9418 (2004), 1407.

20. William Butter, MD, *A Treatise on the Disease Commonly Called Angina Pectoris* (London: J. Johnson, 1791), 10–11.

21. Peter Mere Latham, *Lectures on Subjects Connected with Clinical Medicine, Comprising Diseases of the Heart*, 2nd edn (London: Longman, 1846), 364.

22. Forbes, Tweedie, and Conolly (eds.), *Cyclopaedia*, 82.

23. Caleb Hillier Parry, *An Inquiry into the Symptoms and Causes of the Syncope Anginosa, Commonly Called Angina Pectoris* (London: R. Crutwell, 1799).

24. Parry, *Inquiry*, 86–7.

25. Parry, *Inquiry*, 89.

26. On the competitive nature of eighteenth-century anatomy and professional disputes, see Moore, *The Knife Man*, 64, 83, 109, 112, 192.

27. Pettigrew, 'John Hunter', 119.

28. Pettigrew, 'John Hunter', 119.

29. Jesse Foot, *The Life of John Hunter* (London: T. Becket, 1794), 280.

30. Moore, *The Knife Man*, 45, 125, 346, 347.

31. Lord Holland, *Further Memoirs of the Whig Party 1807–1821* (London: John Murray, 1905), 341–2.

32. John Hunter to Edward Jenner, May 1788, cited in D. Ottley, 'The Life of John Hunter', in *The Works of John Hunter*, ed. Palmer, i. 119.

33. Hunter in *The Works of John Hunter*, ed. Palmer, i. 337.

34. *The Works of John Hunter*, ed. Palmer, i. 131.

35. See Chapter 1.

36. *A Compendium of Anatomy, in which are Described the Figure, Situation, Connection and Uses of the Parts of the Human Body* (London, 1739), 46; Alexander Monro, *The Anatomy of the Humane Bones, to Which Are Added an Anatomical Treatise of the Nerves; An Account of the Reciprocal Motions of the Heart, and a Description of the Human Lacteal Sac and Duct* (Edinburgh: T. and W. Ruddimans; London: J. Osborn and T. Longman, 1732).

37. *Compendium of Anatomy*, 49.

38. Though I am sceptical of the value of retrospective diagnosis, Fleming's interpretation is interesting. See Fleming, *Short History of Cardiology*, 8–9.

39. See Chapter 3.

40. James Hope, *A Treatise on the Diseases of the Heart and the Great Vessels Comprising a New View of the Physiology of the Heart's Action* (London: William Kidd; Edinburgh: Adam Black; Glasgow: T. Atkinson & Co., 1832), 476–77.

41. Hope, *Treatise,* 478.

42. See Chapters 1 and 7.

43. *The Works of John Hunter*, ed. Palmer, i. 45.

44. *The Works of John Hunter*, ed. Palmer, i. 45.

45. Thomas Joseph Pettigrew, John Coakley Lettsom, and Joseph Meredith Toner, *Memoirs of the Life and Writings of the Late John Coakley Lettsom: With a Selection of his Correspondence* (London: Nichols, Son, and Bentley, for Longman, Hurst, Rees, Orme, and Brown, 1817), 295.

46. William Le Fanu, *A Bibliography of Edward Jenner* (London: St Paul's Bibliographies, 1985), 25, citing a letter from Edward Jenner to William Heberden, n.d.

47. *The Works of John Hunter*, ed. Palmer, i. 94.

48. *The Works of John Hunter*, ed. Palmer, i. 45.

49. *The Works of John Hunter*, ed. Palmer, i. 62.

50. Cited in Parry, *Inquiry*, 3–4.

51. Parry, *Inquiry*, 4.

52. Parry, *Inquiry*, 5.

53. See, e.g. Jacalyn M. Duffin, 'Sick Doctors: Bayle and Laennec on their Own Phthisis', *Journal of the History of Medicine and Allied Sciences*, 43 (1998), 165–82.

54. See John Hunter, *Treatise on Venereal Disease*, 2nd edn., trans. and ed. Freeman J. Bumstead (Philadelphia: Blanchard and Lea, 1859), 51, 432–4; Joseph Adams, *A Treatise on the Venereal Disease by John Hunter* (London: Sherwood, Neely and Jones, Soho, 1818); Moore, *Knife Man*, ch. 9.

55. *The Works of John Hunter*, ed. Palmer, i. 119.

56. John Fothergill, 'Farther Account of the Angina Pectoris', *Medical Observations and Inquiries by a Society of Physicians in London*, 5 (1776), 252–8, repr. in *The Works of John Fothergill* (London: Charles Dilly, in the Poultry, 1783–4).

57. See Gerald L. Geison, *Michael Foster and the Cambridge School of Physiology: The Scientific Enterprise in Late Victorian Society* (Princeton: Princeton University Press, 1978), esp. ch. 7.

58. See Elizabeth L. Haigh, 'Vitalism, the Soul and Sensibility: The Physiology of Theophile Bordeu', *Journal of the History of Medicine and Allied Sciences*, 1 (1976), 30–41.

59. Haigh, 'Vitalism', 30. This subject has been touched upon in Chapter 1.

60. John P. Wright and Paul Potter (eds), *Psyche and Soma: Physicians and Metaphysicians on the Mind-Body Problem from Antiquity to Enlightenment* (Oxford: Oxford University Press, 2000), 10. For a more detailed account of these developments, see Fay Bound Alberti, 'Emotions in Early Modern Medical Theory', in ead., *Medicine, Emotion and Disease* (Basingstoke: Palgrave Macmillan, 2006), ch. 1.

61. John Hunter, from *The Works of John Hunter*, ed. Palmer, cited in Guido Cimino and François Duchesneau (eds), *Vitalism: From Haller to the Cell Theory*, proceedings of the Zaragoza Symposium, XIXth International Congress of History of Science, 22–29 Aug. 1993, Biblioteca di Physis 5 (Florence: Leo S. Olschki, 1997), 283.

62. John Hunter, *Works of John Hunter*, ed. Palmer, i. 264.

63. *Works of John Hunter*, ed. Palmer, i. 359.

64. This episode is discussed in Moore, *Knife Man*, 134.

65. John Coakley Lettsom, *The Works of John Fothergill* (London: Charles Dilly, 1873), 268.

66. On the 'non-naturals', see Chapter 1.

67. Bound Alberti, 'Emotions in the Early Modern Medical Tradition', 5.

68. On angina pectoris and men over 40 years of age, see Parry, *Inquiry*, 41.

69. Parry, *Inquiry*, 158.

70. Parry, *Inquiry*, 158.

CHAPTER 3

1. Matthew Baillie, *The Morbid Anatomy of Some of the Most Important Parts of the Human Body* (London: J. Johnson and G. Nicol, 1793; facs. repr. Birmingham, AL: Classics of Medicine Library, 1986), 2.

2. René T. H. Laennec, *A Treatise on the Diseases of the Chest and on Mediate Auscultation*, trans. John Forbes, 2nd edn (London: Thomas and George Underwood, 1827), 549.

3. Baillie, *Morbid Anatomy*, 2.

4. Laennec, *Treatise*, 549.

5. Russell C. Maulitz, 'The Pathological Tradition', in William F. Bynum and Roy Porter (eds), *Companion Encyclopaedia of the History of Medicine* (London: Routledge, 1993), 169–91; Maulitz, *Morbid Appearances: The Anatomy of Pathology in the Early Nineteenth Century* (Cambridge and New York: Cambridge University Press, 1997). On shifting perspectives in the Western scientific and medical tradition more generally, see John V. Pickstone, *Ways of Knowing: A New History of Science, Technology and Medicine* (Manchester: Manchester University Press, 2000).

6. See Russell C. Maulitz, 'Anatomie et anatomo-clinique', in D. Lecourt (ed.), *Dictionaire de la pensée médicale* (Paris: PUF, 2004), 47–51.

7. Peter Fleming, *A Short History of Cardiology* (Amsterdam: Rodopi, 1997), 8–9.

8. See John Harley Warner, *Against the Spirit of System: The French Impulse in Nineteenth-Century American Medicine* (Baltimore, MD: Johns Hopkins University Press, 2003).

9. Lorraine Daston and Peter Galison, *Objectivity* (New York: Zone, 2007).

10. Baillie, *Morbid Anatomy*.

11. Karl Freiherr von Rokitansky, *Handbuch der Pathologischen Anatomie* (3 vols.; Vienna: Braumiller and Seidel, 1842–6). See Alvin E. Rodin, *The Influence of Matthew Baillie's Morbid Anatomy: Biography, Evaluation and Reprint* (Springfield, IL: Charles C. Thomas, 1973), pp. v, 5.

12. Rodin, *Influence*, 15–16.

13. *A Compendium of Anatomy, in which are Described the Figure, Situation, Connection and Uses of the Parts of the Human Body* (London, 1739), 49; this work was published with Alexander Monro, *The Anatomy of the Humane Bones, to which are added an Anatomical Treatise of the Nerves; An Account of the Reciprocal Motions of the Heart, and a Description of the Human Lacteal Sac and Duct* (Edinburgh: T. and W, Ruddimans; London: J. Osborn and T. Longman, 1732).

14. Jean-Baptiste de Senac, *Traité de la structure du cœur, de son action, et de ses maladies* (Paris: J. Vincent, 1749). This work was reproduced posthumously as the shortened version *Traité des maladies du cœur* (1781).

15. Fleming, *Short History of Cardiology*, 7.

16. H. A. Snellen, *History of Cardiology* (Rotterdam: Donker Academic Publications, 1984), 26–7.

17. For an introduction, see Gerard Jorland, Annick Opinel, and George Weisz, *Body Counts: Medical Quantification in Historical and Sociological Perspective/La Quantification Medicale, Perspective Historiques et Sociologiques* (Montreal and Ithaca, NY: McGill-Queen's University Press, 2005).

18. Fleming, *Short History of Cardiology*, 106; George E. Burch, 'The History of Vectorcardiography', John Burnett, 'The Origins of the Electrocardiograph as a Clinical Instrument', and Arthur Hollman, 'The History of Bungle Branch Block', in William F. Bynum, Christopher Lawrence, and Vivian Nutton (eds), *The Emergence of Modern Cardiology* (London: Wellcome Institute for the History of Medicine, 1985), 103–31; 53–76; 82–102.

19. See Robert G. Frank, 'The Tell-Tale Heart: Physiological Instruments, Graphic Methods and Clinical Hopes, 1854–1914', in William Coleman and Frederick L. Holmes (eds), *The Investigative Enterprise: Experimental Physiology in Nineteenth-Century Medicine* (Berkeley and Los Angeles: University of California Press, 1988), 211–90.

20. Leopold Auenbrugger, *Inventum Novum ex Percussione Thoracis Humani Ut Signio Abstrusos Interni Pectoris Morbos Detegendi* (Vindobonae : Joannis Thomae Trattner, 1761); facs. repr. as *Inventum Novum, a facsimile of the first edition with Corvisart's translation (1808), Forbes's English translation (1824), Ungar's German translation (1843); edited with a biographical account by Max Neuburger* (London: Dawsons of Pall Mall, 1966); Laennec, *A Treatise on the Diseases of the Chest*; Fleming, *Short History of Cardiology*, 75.

21. Leopold Auenbrugger, *Inventum Novum*.

22. See S. W. F. Holloway, 'Medical Education in England, 1830–1858: A Sociological Analysis', *History*, 49 (1964), 299–324.

23. For a contextual introduction to the work of Corvisart, see Charles F. Wooley, *The Irritable Heart of Soldiers and the Origins of Anglo-American Cardiology: The US Civil War (1861) to World War I (1918)* (Aldershot: Ashgate, 2002), 88–90.

24. Jean Nicolas Corvisart, *An Essay on the Organic Diseases and Lesions of the Heart and Great Vessels*, trans. Jacob Gates (Boston, 1812; facs. repr. New York: Hafner, 1962), pp. vii, 279. The *Essay* was first published by one of Corvisart's students on the basis of his lectures.

25. Corvisart, *Essay*, p. vii.

26. Corvisart, *Essay*, 17.

27. Corvisart, *Essay*, 17.

28. Corvisart, *Essay*, 20.

29. It was commonplace to regard emotions as 'moral affections' by the mid-nineteenth century. For another example of reference to 'moral affection, emotion or passion' being combined, see Silvester Graham, *Lectures on the Science of Human Life* (London: Horsell and Shirref, 1854), 226.

30. Corvisart, *Essay*, 275.

31. Corvisart, *Essay*, 277.

32. Corvisart, *Essay*, 277.

33. Corvisart, *Essay*, 277-8.
34. Corvisart, *Essay*, 278.
35. Corvisart, *Essay*, 282.
36. Corvisart, *Essay*, 282.
37. Corvisart, *Essay*, 289.
38. Corvisart, *Essay*, 290.
39. Corvisart, *Essay*, 121-2.
40. Corvisart, *Essay*, 123-4.
41. Corvisart, *Essay*, 117.
42. Dickinson W. Richards, Foreword, in Corvisart, *Essay*, n.p.
43. For an excellent introduction, see Jacalyn Duffin, *To See with a Better Eye: A Life of R. T. H. Laennec* (Princeton: Princeton University Press, 1998).
44. Duffin, *To See with a Better Eye*, 131.
45. Duffin, *To See with a Better Eye*, 151.
46. Duffin, *To See with a Better Eye*, 154-5.
47. Duffin, *To See with a Better Eye*, 177.
48. See the discussion in Nancy G. Siraisi, 'The Music of Pulse in the Writings of Italian Academic Physicians (Fourteenth and Fifteenth Centuries)', *Speculum*, 50 (1975), 689-710.
49. See Berndt Lüderitz, *History of the Disorders of Cardiac Rhythm*, 3rd edn (Armonk, NY: Futura, 2002).
50. For an excellent survey of Chinese medicine and the pulses, see Shigehisa Kuriyama, *The Expressiveness of the Body and the Divergence of Greek and Chinese Medicine* (New York: Zone Books, 1999).
51. Lüderitz, *Disorders of Cardiac Rhythm*, 3.
52. Fleming, *Short History of Cardiology*, 21.
53. Sir James Mackenzie, *The Study of the Pulse: Arterial, Venous, and Hepatic, and of the Movements of the Heart* (Edinburgh and London: Young J. Pentland, 1902).
54. Peter Mere Latham, 'General Remarks on the Practice of Medicine: The Heart and its Affections, not Organic', *British Medical Journal*, 1 (1863), 3.
55. See the examples given by Fleming, *Short History of Cardiology*, 109.
56. Frank, 'The Tell-Tale Heart'.
57. Thomas Henry Huxley, 'Biological Apparatus', in South Kensington Museum, *Handbook to the Special Loan Collection of Scientific Apparatus* (London; Chapman and Hall, [1876]), 312-26, at 321.
58. Technological determinism is apparent in Frank's otherwise wide-ranging survey, and in Fleming, *Short History of Cardiology*, 106. For more on the reception of innovative diagnostic techniques, see Chapter 7.
59. In Italy, Antonio Testa, *Della malattie del Cuore* (1811) and in Germany, Friedrich Kreysig, *Die Krankheiten des Herzens* (1814-1817), 3 vols. See Fleming, *Short History of Cardiology*, 34.

60. John Collins Warren, *Cases of Organic Diseases of the Heart. With Dissections and Some Remarks Intended to Point out the Distinctive Symptoms of these Diseases* (Boston: Thomas B. Wait and Company, 1809; Microfiche, New York: Readex Corporation, 1990, 2 microfiches, Second Series, no. 19138); Allan Burns, *Observations on Some of the Most Frequent and Important Diseases of the Heart: On Aneurism of the Thoracic Aorta; On Preternatural Pulsation in the Epigastric Region* (Edinburgh: Muirhead, 1809).

61. Fleming, *Short History of Cardiology*, 32-3.

62. James Hope, *A Treatise on the Diseases of the Heart and Great Vessels, Comprising A New View of the Physiology of the Heart's Action, According to Which the Physical Signs are Explained* (London: W. Kidd; Edinburgh: Adam Black; Glasgow: T. Atkinson & Co., 1832). For an introduction to Hope's life and career, see W. A. Greenhill, 'Hope, James (1801-1841)', rev. Louis J. Acierno, *Oxford Dictionary of National Biography* (*DNB*) (Oxford: Oxford University Press, 2004).

63. Hope, *Treatise*, 479.

64. Walter Hayle Walshe, *A Practical Treatise on the Diseases of the Lungs, and Heart* (London: Taylor, Walton and Maberly, 1851); subsequently published as *A Practical Treatise on the Diseases of the Heart and Great Vessels, Including and Principles of their Physical Diagnosis*, 4th edn. (London: Smith, Elder, 1873).

65. The following information is taken from Iain Milne, 'Bramwell, Sir Byrom (1847-1931)', *DNB*.

66. Byrom Bramwell, *Diseases of the Heart and Thoracic Aorta* (Edinburgh: Y. J. Pentland, 1884), 1.

67. Bramwell, *Diseases of the Heart*, 2.

68. Bramwell, *Diseases of the Heart*, 9. Foster's *Textbook* was translated into Russian, German, and Italian, and remained in print through six editions, and part of a seventh. This is a remarkable statistic, given that English physiology had previously been of limited influence. See Terrie M. Romano, 'Foster, Sir Michael (1836-1907)', *DNB*.

69. Bramwell, *Diseases of the Heart*, 11. Walter Holbrook Gaskell's publications included 'On the Tonicity of the Heart and Arteries', *Proceedings of the Royal Society of London*, 30 (1879), 225-7.

70. Bramwell, *Diseases of the Heart*, 17.

71. Bramwell, *Diseases of the Heart*, 28.

72. Exophthalmic goitre is a condition caused by excessive production of thyroid hormone and characterized by an enlarged thyroid gland, protrusion of the eyeballs, a rapid heartbeat, and nervous excitability. Also known as 'Graves' Disease'.

73. Bramwell, *Diseases of the Heart*, 47.

74. Bramwell, *Diseases of the Heart*, 47.

75. Bramwell, *Diseases of the Heart*, 62.

76. Blacksmiths and those in strenuous activities were more likely to get aortic lesions and aneurism, and soldiers and sailors were more prone to arterial disease because of social and environmental factors that included 'strain' (physical and mental, presumably), as well as excessive exposure to syphilis and alcohol. Bramwell, *Diseases of the Heart*, 62–3.

77. Christopher Lawrence, 'Moderns and Ancients: The "New Cardiology" in Britain, 1880–1930', in Bynum et al. (eds), *The Emergence of Modern Cardiology*, 5. William Osler, *The Principles and Practice of Medicine* (Edinburgh and London: Young J. Pentland, 1892).

78. Lawrence, 'Moderns and Ancients', 5.

79. Lawrence, 'Moderns and Ancients', 6.

80. Lawrence, 'Moderns and Ancients', 6.

81. William Osler, *The Principles and Practice of Medicine*, 3rd edn (Edinburgh and London: Young J. Pentland, 1898), 5. All following citations are from this edition.

82. Osler, *The Principles and Practice*, 5.

83. Osler, *The Principles and Practice*, 692.

84. Osler, *The Principles and Practice*, 755.

85. Osler, *The Principles and Practice*, 756.

86. Osler, *The Principles and Practice*, 756. See also Joel D. Howell, ' "Soldier's Heart": The Redefinition of Heart Disease and Specialty Formation in Early Twentieth-Century Great Britain', in Bynum et al. (eds), *The Emergence of Modern Cardiology*, 34–52.

87. Osler, *The Principles and Practice*, 757.

88. Wooley, *Irritable Heart*.

89. Jacob Mendes Da Costa, 'On Irritable Heart: A Clinical Study of a Form of Functional Cardiac Disorder and its Consequences', *American Journal of the Medical Sciences*, 61 (1871), 17–52.

90. Sir (Thomas) Clifford Allbutt (ed.), *A System of Medicine by Many Writers* (London: Macmillan, 1896–9). Allbutt also published *The Effects of Overwork and Strain on the Heart and Great Blood-Vessels* ([St George's Hospital Reports; London: J. and A. Churchill?], 1870). For a biographical introduction see H. D. Rolleston, 'Allbutt, Sir (Thomas) Clifford (1836–1925)', rev. Alexander G. Bearn, *DNB*.

91. Thomas Lewis, *The Soldier's Heart and the Effort Syndrome* (London: Shaw, 1918). For a biographical note, see Arthur Hollman, 'Lewis, Sir Thomas (1881–1945)' *DNB*.

92. See, e.g., Graham Dawson, *Soldier Heroes: British Adventure, Empire and the Imagining of Masculinities* (London and New York: Routledge, 1994).

93. Allan Young, *The Harmony of Illusions: Inventing Post-Traumatic Stress Disorder* (Princeton: Princeton University Press, 1997), 52.

94. Osler, *The Principles and Practice*, 766.

95. Osler, *The Principles and Practice*, 111.

96. For an introduction to the history of hysteria, see Ilza Veith, *Hysteria: The History of a Disease* (Chicago: University of Chicago Press, 1970).

97. Osler, *The Principles and Practice*, 1119.

98. Osler, *The Principles and Practice*, 1122.

99. Osler, *The Principles and Practice*, 760–1; M. R. Wilkins, M. J. Kendall, and O. L. Wade, 'William Withering and Digitalis', *British Medical Journal*, 290 (1985), 7–8.

100. Bramwell, *Diseases of the Heart*, 48.

101. Bramwell, *Diseases of the Heart*, 50.

102. Bramwell, *Diseases of the Heart*, 51–2.

103. Fleming, *Short History of Cardiology*, 143. The most wide-ranging and insightful essay on this subject remains Lawrence, 'Moderns and Ancients', 1–33.

104. Alexander Mair, *Sir James Mackenzie MD: 1853–1925: General Practitioner*, 2nd edn (London: Royal College of General Practitioners, 1986), and Arthur Hollman, *Sir Thomas Lewis: Pioneer Cardiologist and Clinical Scientist* (London: Springer, 1997).

105. Fleming, *Short History of Cardiology*, 144.

106. Fleming, *Short History of Cardiology*, 145.

107. On the scale of technological and scientific investigation more generally, see William Coleman and Frederic L. Holmes, 'Introduction', in eid. (eds), *The Investigative Enterprise: Experimental Physiology in Nineteenth-Century Medicine* (Berkeley and Los Angeles, and London: University of California Press, 1988), 3–4.

108. Coleman and Holmes, 'Introduction', 4. See also Kathryn M. Olesko, 'Commentary: On Institutes, Investigations, and Scientific Training', in Coleman and Holmes (eds), *The Investigative Enterprise*, 295–332.

109. For an introduction, see Thomas Willis, *Dr Willis's Practice of Physick* (London: T. Dring, C. Harper, and J. Leigh, 1684).

110. See Lüderitz, *Disorders of Cardiac Rhythm*.

111. See Arthur J. Moss, 'History of the Origin of the Heart Beat: "On the Innervation of the Heart"', *Annals of Noninvasive Electrocardiology*, 5/3 (2000), 290–1.

112. Walter H. Gaskell, 'On the Structure, Distribution, and Function of the Nerves which Innervate the Visceral and Vascular Systems', *Journal of Physiology*, 7 (1886), 1–80, at 40–1.

113. See W. F. Windle, 'John Newport Langley (1852–1925)', in W. Haymaker and F. Schiller (eds), *The Founders of Neurology*, 2nd edn (Springfield, IL: Charles C. Thomas, 1970), 289–92 and the discussion in Stanley Finger, *Origins of Neuroscience: A History of Explorations into Brain Function* (New York and Oxford: Oxford University Press, 1994). The autonomic nervous system is today believed to control the involuntary actions of the body, including the heartbeat, respiration, and sweating.

114. See Chapter 7.

115. Otniel E. Dror, 'Fear and Loathing in the Laboratory and Clinic', in Fay Bound Alberti (ed.), *Medicine, Emotion and Disease, 1700–1950* (Basingstoke: Palgrave Macmillan, 2006), 125–43, and Dror, 'The Scientific Image of Emotion: Experience and Technologies of Inscription', *Configurations*, 7/3 (1999), 355–401.

116. Otniel Dror, 'The Affect of Experiment: The Turn to Emotions in Anglo-American Physiology, 1900–1940', *Isis*, 90 (1999), 205–37, at 207.

117. See Herman A. Snellen, *E. J. Marey and Cardiology* (Rotterdam: Kooyker, 1980), 127–46, Marta Braun, *Picturing Time: The Work of Étienne-Jules Marey (1830–1904)* (Chicago: Chicago University Press, 1992), and Christopher Lawrence, 'Physiological Apparatus in the Wellcome Museum. 1. The Marey Sphygmograph', *Medical History*, 22 (1978), 196–200.

118. See Fleming, *Short History of Cardiology*, 115, and Lawrence, 'Physiological Apparatus'.

119. Frank, 'The Telltale Heart', 212.

120. 'Physicians and Physicists', *Lancet*, 2 (1865), 599.

121. *La Méthode graphique dans les sciences expérimentales et principalement en physiologie et en medecine* (1878), cited in Frank, 'The Telltale Heart', 218.

122. See Frederic L. Holmes and Kathryn M. Olesko, 'The Images of Precision: Helmholtz and the Graphical Method in Physiology', in M. Norton Wise (ed.), *The Values of Precision* (Princeton: Princeton University Press, 1995), 198–222.

123. Christopher Lawrence, 'Incommunicable Knowledge: Science, Technology and the Clinical Art in Britain 1850–1914', *Journal of Contemporary History*, 20 (1985), 503–20, at 504.

124. Cited in Lawrence, 'Incommunicable Knowledge', 505. The quotation is used to similar ends in an article with a shared conclusion to that by Lawrence: Hughes Evans, 'Losing Touch: The Controversy over the Introduction of Blood Pressure Instruments into Medicine', *Technology and Culture*, 34/4: Special Issue: 'Biomedical and Behavioural Technology' (1993), 784–807.

125. Lawrence, 'Incommunicable Knowledge', 506.

126. Evans, 'Losing Touch'. See also Joel D. Howell, 'Early Perceptions of the Electrocardiogram: From Arrhythmia to Infarction', *Bulletin of the History of Medicine*, 58 (1984), 83–98.

127. Lawrence certainly thinks that it was an important aspect of defining the English gentleman physician: 'Incommunicable Knowledge', 512, 517.

128. See the discussion in Dror, 'Affect of Experiment', 211. For a traditional introduction of the context in which Bernard was working, see Hebbel H. Hoff and Roger Guillermin, 'Claude Bernard and the Vasomotor System', in Francisco Grande and Maurice B. Visscher (eds),

Claude Bernard and Experimental Medicine (Cambridge, MA: Schenkman Publishing Company, 1967).

129. See Steve Sturdy, 'Sanderson, Sir John Scott Burdon, baronet (1828–1905)', *DNB*. For a discussion of the politicking surrounding its adoption in British clinical medicine, see Frank, 'The Telltale Heart', 219–20.

130. Dror, 'Affect of Experiment', 211.

131. Dror, 'Affect of Experiment', 211.

132. George Rosen, 'Changing Attitudes of the Medical Profession to Specialization', *Bulletin of the History of Medicine*, 12 (1942), 343–54; J. D. Howell, ' "Soldier's Heart": The Redefinition of Heart Disease and Speciality Formation in Early Twentieth-Century Great Britain', and Lawrence, 'Moderns and Ancients', in Bynum et al., *The Emergence of Modern Cardiology*, 34–52; 1–33.

133. Lawrence, 'Moderns and Ancients', 1, 10–11; Maurice Campbell, 'The National Heart Hospital, 1857–1957', *British Heart Journal*, 20 (1958), 137–9.

134. See Bynum et al. (eds), *The Emergence of Modern Cardiology*, introduction, p. ix.

135. A. Mair, *Sir James Mackenzie MD 1853–1925: General Practitioner* (London: Royal College of General Practice, 1986), 107.

136. Dror, 'Affect of Experiment', 218. See Elin L. Wolfe, *Walter B. Cannon: Science and Society* (Boston: Boston Medical Library in the Francis A. Countway Library of Medicine, 2000).

137. Duffin, *To See with a Better Eye*, 237.

CHAPTER 4

1. Peter Mere Latham, *Lectures on Subjects Connected with Clinical Medicine, Comprising Diseases of the Heart* (2 vols, London: Longman, Brown, Green and Longmans, 1845), i. 376.

2. I have used the case of Thomas Arnold to explore more generally the relationship between functional and structural disease in Fay Bound Alberti, 'Angina Pectoris and the Arnolds: Emotions and Heart Disease in Nineteenth-Century Culture', *Medical History*, 52 (2008), 221–36.

3. On William Charles Lake (1817–97), later dean of Durham, and on Thomas Arnold (1795–1842), see M. C. Curthoys, 'Lake, William Charles (1817–1897)' and A. J. H. Reeve, 'Arnold, Thomas (1795–1842)', *Oxford Dictionary of National Biography* (*DNB*) (Oxford: Oxford University Press, 2004).

4. Arthur Penrhyn Stanley, *Life of Thomas Arnold, D.D. Headmaster of Rugby* (1844; repr. London: John Murray, 1904), 655.

5. On Matthew Arnold, see Stefan Collini, 'Arnold, Matthew (1822–1888)', *DNB*.

6. Park Honan, *Matthew Arnold: A Life* (London: Weidenfeld and Nicolson, 1981), reviewed by Miriam Allott in *The Modern Language Review*, 79 (1984), 164–7.

7. Kirstie Blair, *Victorian Poetry and the Culture of the Heart* (Oxford: Oxford University Press, 2006); Christopher Lawrence, ' "Definite and Material": Coronary Thrombosis and Cardiologists in the 1920s', in Charles E. Rosenberg and Janet Golden (eds), *Framing Disease: Studies in Cultural History* (New Brunswick, NJ: Rutgers University Press, 1992), 51–82; id., *Rockefeller Money The Laboratory and Medicine in Edinburgh 1919–1930* (Rochester, NY: University of Rochester Press, 2005); Charles F. Wooley, *The Irritable Heart of Soldiers and the Origins of Anglo-American Cardiology: the US Civil War (1861) to World War I (1918)* (Aldershot: Ashgate, 2002).

8. See Wooley, *Irritable Heart*, 65. The explicit association of 'functional' disease with emotional excess and neuroses was not made until the late nineteenth century. See the discussion in Chapter 8.

9. An important exception is the work of Christopher Lawrence. See his review of Christopher Crenner, *Private Practice: In the Twentieth-Century Medical Office of Dr Richard Cabot* (Baltimore: Johns Hopkins University Press, 2005), in 'Dr Cabot and Mr Hyde', *Medical History*, 50 (2006), 247, and Christopher Lawrence and George Weisz (eds), *Greater than the Parts: Holism in Biomedicine, 1920–1950* (New York and Oxford: Oxford University Press, 1998).

10. See Wooley, *Irritable Heart*, 67.

11. Blair, *Victorian Poetry*, 30.

12. Charles E. Rosenberg, 'Body and Mind in Nineteenth-Century Medicine: Some Clinical Origins of the Neurosis Construct', *Bulletin for the History of Medicine*, 63 (1989), 185–97; John Harley Warner, *The Therapeutic Perspective: Medical Practice, Knowledge, and Identity in America, 1820–1885* (Cambridge, MA, and London: Harvard University Press, 1986).

13. Robert G. Frank, 'The Tell-Tale Heart: Physiological Instruments, Graphic Methods and Clinical Hopes, 1854–1914', in William Coleman and Frederick L. Holmes (eds), *The Investigative Enterprise: Experimental Physiology in Nineteenth-Century Medicine* (Berkeley and Los Angeles: University of California Press, 1988), 211–90. For a recent critique of the traditional story of nineteenth-century medical development as reductionist and technologically defined, see Lawrence and Weisz, 'Medical Holism: The Context', in eid. (eds), *Greater than the Parts*.

14. Peter Mere Latham, *Lectures on Subjects Connected with Clinical Medicine, Comprising Diseases of the Heart* (2 vols; London: Longman, Brown, Green and Longmans, 1845), i. 373; Stanley, *Life*, 655.

15. Latham, *Lectures*, 375.

16. See Christopher Lawrence, 'Moderns and Ancients: The "New Cardiology" in Britain, 1880–1930', in William F. Bynum, Christopher Lawrence, and Vivian Nutton (eds), *The Emergence of Modern Cardiology* (London: Wellcome Institute for the History of Medicine, 1985).

17. *The Times*, Monday, 25 Mar. 1872, p. 7; issue 27333; col. C. For another contemporary account of the rise of heart disease, see Herbert Davies, *Lectures on the Physical Diagnosis of the Diseases of the Lungs and Heart* (London: John Churchill, 1851), 1–2.

18. This provides an interesting contrast to the literary depiction of heart disease, which at least one scholar has identified as gendered towards women. Blair, *Victorian Poetry*, 37. For an account of the gendering of heart disease today, see Gerdi Weidner et al. (eds), *Heart Disease: Environment, Stress and Gender* (NATO Science Series, Series I, vol. 327; Budapest, 2000).

19. *The Times*, p. 7.

20. Jean Nicolas Corvisart, *An Essay on the Organic Diseases and Lesions of the Heart and Great Vessels*, trans. Jacob Gates (1806; facs. repr. Hafner: New York, 1962), 26, 28.

21. Except where noted, the following account is derived from Stanley, *Life*, 655–6. See also Meriol Trevor, *The Arnolds: Thomas Arnold and his Family* (London, Bodley Head, 1973), 45.

22. Trevor, *Arnolds*, 45.

23. Stephen Stansfeld and Michael Marmot (eds), *Stress and the Heart: Psychosocial Pathways to Coronary Heart Disease* (London: BMJ Books, 2002). As noted above it is important to be aware of the historical specificity of concepts like 'stress'. For an introduction, see Rhodri Hayward, 'Stress', *Lancet*, 365/9476 (2005), 2001.

24. Not in the Freudian sense in which civilization, or its culture, inhibits man's instinctual drives, which can (and perhaps must) result in guilt and non-fulfilment. See Sigmund Freud, *Unbehagen in der Kultur* (1930), repr. in *Civilization and its Discontents*, trans. David McLintock (London: Penguin, 2004). The term 'disease of civilization' has since been taken up to indicate 'lifestyle' diseases. See Mark Harrison and Michael Worboys (eds), *A Disease of Civilisation: Tuberculosis in Britain, Africa and India, 1900–39* (London: Routledge, 1997). For a contemporary perspective, see Benjamin Richardson, *Diseases of Modern Life* (London: Macmillan, 1876), 120.

25. Stanley, *Life*, 657.

26. Wooley, *Irritable Heart*, 65.

27. Wooley, *Irritable Heart*, 66.

28. Allan Burns, *Observations on Some of the Most Frequent and Important Diseases of the Heart* (1809; repr. New York: Hafner, 1964), 140.

29. Wooley, *Irritable Heart*, 95.

30. See, e.g. Byrom Bramwell, *Diseases of the Heart and Thoracic Aorta* (Edinburgh: Y. J. Pentland, 1884), 659.

31. Latham, *Lectures*, 376.
32. On Joseph Hodgson, see G. T. Bettany, 'Hodgson, Joseph (1788–1869)', rev. Peter R. Fleming, *DNB*. Thanks to Jonathan Reinarz for valuable discussions of the significance of Hodgson in relation to national and local medical practice.
33. Latham, *Lectures*, 377.
34. Latham, *Lectures*, 377.
35. Latham, *Lectures*, 379.
36. Latham, *Lectures*, 361.
37. Latham, *Lectures*, 362.
38. Latham, *Lectures*, 361. This theme was addressed by many physicians in the nineteenth century and by William Osler in the twentieth century.
39. Latham, *Lectures*, 386.
40. Latham, *Lectures*, 385.
41. Latham, *Lectures*, 405.
42. Latham, *Lectures*, 405.
43. Latham, *Lectures*, 405.
44. For a recent discussion and critique of the latter, see P. Claesa, 'A True Story of "Organ-Based Medicine", Or, When Did We Stop Thinking?' *European Journal of Internal Medicine*, 18 (2007), 173–4.
45. For more discussion of Latham's practice and holism, see chapter 5, above.
46. Latham, *Lectures*, 408.
47. Matthew Arnold, *Selected Letters of Matthew Arnold*, ed. Clinton Machann and Forrest D. Burt (Macmillan: London, 1993), 283.
48. Latham, *Lectures*, 374.
49. See Lawrence and Weisz, 'Medical Holism'.
50. See Fay Bound Alberti, 'The Emotional Heart' in James Peto (ed.), *The Heart* (London: Yale University Press, 2007), pp. 125–42.

CHAPTER 5

1. Peter Mere Latham, Medical Case Book, 1 Nov. 1840–31 Dec. 1844, Wellcome Library, WMS.3177. A stipple engraving of Latham is provided in Fig. 5.
2. This is not Sir John Forbes, the physician, who would have been 53 in 1840. See R. A. L. Agnew, 'Forbes, Sir John (1787–1861)', *Oxford Dictionary of National Biography* (*DNB*) (Oxford: Oxford University Press, 2004).
3. The development of Ventnor as a therapeutic centre is discussed in more detail later in this chapter. I am grateful to Alan Champion for sharing with me his work on the history of Ventnor and its doctors.
4. A. Courtney, 'Treatment of Epilepsy: With the Digitalis and Polypodium', *Lancet*, 17/243 (1831), 45–6; 'Digitalis and Cardiac Work in Heart Disease', *Lancet*, 232/6011 (1938), 1128.

5. See James Currie, *Medical Reports, on the Effects of Water, Cold and Warm, as a Remedy in Fever and Other Diseases, Whether Applied to the Surface of the Body or Used Internally* (2 vols, London: M'Creery for Cadell & Davies, 1804), vol. i; Samuel Weeding, *The Wet Sheet. Addressed to the Medical Men of England. Cases, Illustrative of the Powerful and Curative Effect of the Wet Sheet, With the External and Internal Application of Water in Acute Diseases, as Occurred in Private Practice* (London: Churchill, 1843). Thanks to Alan Champion for this reference.

6. For a biography of Latham, see Peter R. Fleming, 'Latham, Peter Mere (1789–1875)', *DNB*.

7. See, e.g. *An Account of the Disease Lately Prevalent at the General Penitentiary* (London: Thomas and George Underwood, 1825); now available on microfiche as *The Nineteenth Century Title No. N.1.1.11738 General Collection* (Cambridge: Chadwyck-Healy Ltd, 1999).

8. Peter Mere Latham, *Lectures on Subjects Connected with Clinical Medicine, Comprising Diseases of the Heart* (London: Longman, Rees, Orme, Brown, Green, and Longman, 1836).

9. Latham, *Lectures*, and 'General Remarks on the Practice of Medicine: The Heart and its Affections, Not Organic', *British Medical Journal*, 1 (3–24 Jan. 1863), 1–3, 27–30, 53–6, 79–82.

10. James M. Williamson, *Ventnor and the Undercliff in Chronic Pulmonary Diseases* (London: Bailliere, Tindall and Cox, 1884), 1.

11. See Sir James Clark, *The Influence of Climate in the Prevention and Cure of Chronic Diseases, More Particularly of the Chest and Digestive Organs* (London: T. and G. Underwood, 1830).

12. Williamson, 'The Effect of the Climate Medically', in *Ventnor*, 30.

13. Williamson, *Ventnor*, 34.

14. Williamson, *Ventnor*, 32.

15. A description of the proposals and record of the opening of the hospital are given in the *Ventnor Times*, 21 Sept. 1867, n.p., and the *Isle of Wight Observer*, 13 Nov. 1869, n.p.

16. E. W. Laidlaw, *The Story of the Royal National Hospital, Ventnor* (Newport, Isle of Wight: Crossprint, 1990), 7.

17. See Chapters 4 and 6. On the construction of a culture of heart disease, see Kirstie Blair, *Victorian Poetry and the Culture of the Heart* (Oxford: Oxford University Press, 2006).

18. Wellcome Library, WMS.3176 and WMS.3177.

19. See, e.g. Charles E. Rosenberg, 'Body and Mind in Nineteenth-Century Medicine: Some Clinical Origins of the Neurosis Construct', *Bulletin of the History of Medicine*, 63 (1989), 185–97; id., 'Medical Text and Social Context: Explaining William Buchan's *Domestic Medicine*', *Bulletin of the History of Medicine*, 57 (1983), 22–42; Charles E. Rosenberg and Janet Golden (eds), *Framing Disease: Studies in Cultural History* (New

Brunswick, NJ: Rutgers University Press, 1992); John Harley Warner, *Against the Spirit of System: The French Impulse in Nineteenth-Century American Medicine* (Baltimore: Johns Hopkins University Press, 2003); Frank Huisman and John Harley Warner, *Locating Medical History: The Stories and their Meanings* (Baltimore and London: Johns Hopkins University Press, 2006).

20. See J. Dallas, 'The Cullen Consultation Letters', *Proceedings of the Royal College of Physicians of Edinburgh*, 31 (2001), 66–8.

21. It is important to note that even apparently subjective texts are shaped by a series of objectively determined conventions and stylistic practices. On a related theme, see my work on love letters brought before the eighteenth-century ecclesiastical courts in marriage cases: Fay Bound, ' "Writing the Self": Love and the Letter in England, *c*.1660–*c*.1760', *Literature and History*, 11 (2002), 1–19.

22. For a modern example, see Kenneth Rochel de Camargo, Jr, 'The Thought Style of Physicians: Strategies for Keeping up with Medical Knowledge', *Social Studies of Science*, 32 (2002), 827–55.

23. William B. Bean, *Aphorisms from Latham* (Iowa City: Prairie Press [1962]), 54.

24. Bean, *Aphorisms*, 51.

25. Bean, *Aphorisms*, 25.

26. See Chapter 3.

27. Byrom Bramwell, *Diseases of the Heart and Thoracic Aorta* (Edinburgh: Y. J. Pentland, 1884), 57.

28. Bramwell, *Diseases*, 58.

29. Bramwell, *Diseases*, 58.

30. Bean, *Aphorisms*, 72.

31. Bramwell, *Diseases*, 59.

32. Bramwell, *Diseases*, 59.

33. Peter Mere Latham, 'General Remarks', 1–3.

34. For an introduction to the use of metaphor in medical discourses, see Susan Sontag, *AIDS and its Metaphors* (London: Penguin, 1989), and Evelyn Fox Keller, *Refiguring Life: Metaphors of Twentieth-Century Biology* (New York and Chichester: Columbia University Press, 1995).

35. See the discussion in Chapter 3.

36. Latham, 'General Remarks', 3.

37. Latham, *Lectures*, 3.

38. Blair, *Victorian Poetry*.

39. For a discussion of sulphur and its uses in treatment of the skin, see K. S. Leslie, G. W. M. Millington, and N. J. Levell, 'Sulphur and Skin: From Satan to Saddam', *Journal of Cosmetic Dermatology*, 3/2 (2004), 94–8.

40. The history of the tongue as both cultural symbol and medical organ, including its role in diagnosis, is a subject deserving further attention.

41. See Barbara Kirschbaum, *Atlas of Chinese Tongue Diagnosis* (Seattle: Eastland Press, 2000–3).
42. See the discussion in Chapter 4, and Fay Bound Alberti, 'Angina Pectoris and the Arnolds: Emotions and Heart Disease in the Nineteenth Century', *Medical History*, 52 (2008), 221–36.
43. For an introduction to Thomas Watson, see Norman Moore, 'Watson, Sir Thomas, first baronet (1792–1882)', rev. Anita McConnell, *DNB*.
44. Latham, *Lectures*, 407.
45. On the 'non-naturals' and emotion, see Fay Bound Alberti, 'The Emotional Heart: Mind, Body and Soul', in James Peto (ed.), *The Heart* (New Haven and London: Yale University Press, 2007), and ead., 'Introduction: Emotion Theory and Medical History', in Fay Bound Alberti (ed.), *Medicine, Emotion and Disease 1700–1950* (New York and Basingstoke: Palgrave Macmillan, 2006). For a discussion of ancient medical practices in which digestion and excretion are contextualized as bodily processes, see Vivian Nutton, *Ancient Medicine* (London and New York: Routledge, 2004), chs 5 and 9. For the later period, see George S. Rousseau, 'Coleridge's Dreaming Gut: Digestion, Genius, Hypochondria', and other essays, in Christopher E. Forth and Ana Carden-Coyne (eds), *Cultures of the Abdomen: Diet, Digestion and Fat in the Modern World* (Basingstoke: Palgrave Macmillan, 2005).
46. See Roger French and Andrew Wear (eds), *British Medicine in an Age of Reform* (New York: Routledge, 1991), introduction, pp. 5–6.
47. Mary Fissell puts this process relatively early in the nineteenth century. See 'The Disappearance of the Patient's Narrative and the Invention of Hospital Medicine', in French and Wear (eds), *British Medicine*, 92.
48. See *The Collected Works of Dr P. M. Latham with Memoir by Sir Thomas Watson*, ed. for the society by R. Martin, 2 vols (London: New Sydenham Society, 1876–8).
49. Latham, *Collected Works*, ed. Martin, I, pp. xxxvi–xxxvii.
50. Important examples include Rosenberg, 'Body and Mind'; Harley Warner, *Against the Spirit of System*; Christopher Lawrence and George Weisz (eds), *Greater than the Parts: Holism in Biomedicine, 1920–1950* (New York and Oxford: Oxford University Press, 1998).
51. Bound Alberti, 'The Heart of Emotions', and ead., 'Emotion Theory and Medical History'.

CHAPTER 6

1. *Harriet Martineau's Autobiography*, introduced by Gaby Weiner (3 vols; London, 1877; London: Virago, 1983), ii. 424. Thanks to Sam Alberti, Jo Alberti, and Thomas Dixon for discussing Harriet Martineau and this chapter with me. I would also like to express my gratitude to Barbara Todd, for welcoming me to the Knoll, and fostering still further my interest in Harriet Martineau.

2. Michel Foucault, *Language, Counter-Memory, Practice: Selected Essays and Interviews*, ed. Donald F. Bouchard, trans. Donald F. Bouchard and Sherry Simon (Oxford: Blackwell, 1977).

3. The bibliography on this subject is too vast to be addressed here. By way of introduction, see Louisa Young, *The Book of the Heart* (London: Flamingo, 2002); Emily Jo Sargent, 'The Sacred Heart', in James Peto (ed.), *The Heart* (New Haven and London: Yale University Press, 2007) 102–14.

4. For an introduction to sensibility and sensitivity as literary movements, and to the heartfelt emotions of poetry and prose, see Jerome McGann, *The Poetics of Sensibility: A Revolution in Literary Style* (Oxford: Oxford University Press, 1998); Anne Vincent-Buffault, *A History of Tears: Sensibility and Sentimentality in France*, trans. Teresa Bridgeman (Basingstoke: Macmillan, 1990); Markman Ellis, *The Politics of Sensibility: Race, Commerce, and Gender in the Sentimental Novel* (Cambridge Studies in Romanticism, 12; New York: Cambridge University Press, 1996).

5. Kirstie Blair, *Victorian Poetry and the Culture of the Heart* (Oxford: Oxford University Press, 2006). On identity and its conflicts, see Akhil Gupta and James Ferguson (eds.), *Culture, Power, Place: Explorations in Critical Anthropology* (Durham, NC, and London: Duke University Press, 1997).

6. There is a growing body of writing on Harriet Martineau and her illness, in connection with health studies, sociology, and literary theory. Recent examples include Ellen Annandale, 'Assembling Harriet Martineau's Gender and Health Jigsaw', *Women's Studies International Forum*, 30 (2007), 355–66; Susan F. Bohrer, 'Harriet Martineau: Gender, Liability and Disability', *Nineteenth-Century Contexts*, 25 (2003), 21–37; Maria Frawley, 'Harriet Martineau: Health and Journalism', *Women's Writing*, 9 (2002), 433–44.

7. Examples of the rich Martineau historiography include Caroline Roberts, *The Woman and the Hour: Harriet Martineau and Victorian Ideologies* (Toronto: University of Toronto Press, 2002); Linda H. Peterson, 'Harriet Martineau: Masculine Discourse, Female Sage', *Victorian Sages and Cultural Discourse*, ed. Thaïs Morgan (New Brunswick, NJ: Rutgers University Press, 1990), 171–86; Ann Hobart, 'Harriet Martineau's Political Economy of Everyday Life', *Victorian Studies*, 37/2 (1994), 223–52; Deborah Ann Logan, *The Hour and the Woman: Harriet Martineau's 'Somewhat Remarkable' Life* (Dekalb, IL: Northern Illinois University Press, 2002).

8. Martineau's accounts were subsequently published as a book. See Harriet Martineau, *Miss Martineau's Letters on Mesmerism* (New York: Harper and Brothers, 1845). There is an ample literature on mesmerism and Harriet Martineau, including her relationship with Greenhow, that does not need to be reproduced here. See, e.g. Frawley's discussion in Harriet Martineau, *Life in the Sickroom*, ed. Maria H. Frawley

(Ormskirk: Broadview, 2003), 21, and the reproduction of Greenhow's report, p. 187, app. C.

9. T. Greenhow, *Facts in the Case of Miss H—M—* (London, 1845).

10. Martineau, *Autobiography*, ii. 198.

11. See Barbara Todd, *Harriet Martineau at Ambleside, with 'A Year at Ambleside' by Harriet Martineau* (Carlisle: Bookcase, 2002).

12. Martineau, *Autobiography*, ii. 431.

13. Cited in Gayle Graham Yates (ed.), *Harriet Martineau on Women* (New Brunswick, NJ: Rutgers University Press, 1985), 35–49.

14. Bohrer, 'Harriet Martineau: Gender, Liability and Disability'; Frawley, 'Harriet Martineau: Health and Journalism'.

15. Alison Winter, 'Harriet Martineau and the Reform of the Invalid in Victorian Britain', *Historical Journal*, 38 (1995), 597–616. See also ead., *Mesmerized: Powers of Mind in Victorian Britain* (London: University of Chicago Press, 1998); Roger Cooter, 'Dichotomy and Denial: Mesmerism, Medicine and Harriet Martineau', in Marina Benjamin (ed.), *Science and Sensibility: Gender and Scientific Enquiry, 1780–1945* (Oxford: Basil Blackwell, 1991), 156–7. See also Maria Frawley, ' "A Prisoner to the Couch": Harriet Martineau, Invalidism and Self-Representation', in David T. Mitchell and Sharon L. Snyder (eds), *The Body and Physical Difference: Discourses on Disability* (Ann Arbor: University of Michigan Press, 1997), 174–88; Trev Lynn Broughton, 'Making the Most of Martyrdom: Harriet Martineau, Autobiography and Death', *Literature and History*, 2 (1993), 24–45.

16. R. K. Webb, *Harriet Martineau: A Radical Victorian* (New York: Columbia University Press, 1960), 7.

17. Anka Ryall, 'Medical Body and Lived Experience: The Case of Harriet Martineau', *Mosaic*, 33 (2000), 35–53; Cooter, 'Dichotomy and Denial'. For an insightful analysis of the 'medical' body and the theatrics of treatment, see Athena Vrettos, *Somatic Fictions: Imagining Illness in Victorian Culture* (Stanford, CA: Stanford University Press, 1995), 4–6.

18. The best example of this is Blair, *Victorian Poetry*.

19. Peter Mere Latham, *Lectures on Subjects Connected with Clinical Medicine, Comprising Diseases of the Heart* (London: Longman, Rees, Orme, Brown, Green, and Longman, 1836); Peter R. Fleming, 'Latham, Peter Mere (1789–1875)', *Oxford Dictionary of National Biography (DNB)* (Oxford: Oxford University Press, 2004).

20. Ryall, 'Medical Body and Lived Experience'.

21. Norman Moore, 'Watson, Sir Thomas, first baronet (1792–1882)', rev. Anita McConnell, *DNB*.

22. See Vera Wheatley, *The Life and Work of Harriet Martineau* (London: Secker and Warburg, 1957), 347.

23. Martineau, *Autobiography*, ii. 424.

24. Martineau, *Autobiography*, ii. 431. For an introduction to the importance of autobiography to nineteenth-century women, and the shaping of autobiography narratives, see Valerie Sanders, *The Private Lives of Victorian Women: Autobiography in Nineteenth-Century England* (New York and London: Harvester Wheatsheaf, 1989); Elaine Showalter, *A Literature of their Own: British Women Novelists from Brontë to Lessing* (Princeton: Princeton University Press, 1977; London: Virago, 1978); Timothy Pelatson, 'Life Writing', in Herbert F. Tucker (ed.), *A Companion to Victorian Literature and Culture* (Oxford: Blackwell, 1999), 356–72.

25. See E. E. Rea, 'Matthew Arnold on Education: Unpublished Letters to Harriet Martineau', *The Yearbook of English Studies*, 2 (1972), 181–91, at 181. Published letter collections of Harriet Martineau include Deborah Anna Logan (ed.), *The Collected Letters of Harriet Martineau* (5 vols; London: Pickering & Chatto, 2007); Elizabeth Sanders Arbuckle (ed.), *Harriet Martineau's Letters to Fanny Wedgwood* (Stanford, CA: Stanford University Press, 1983); and Valerie Sanders (ed.), *Harriet Martineau: Selected Letters* (Oxford: Oxford University Press; New York: Oxford University Press, 1990).

26. Birmingham University Library, Special Collections, letter from Latham to Martineau, HM 539.

27. 'Battley's Laudanum' or 'Battley's Sedative Solution' was a sedative composed of opium dissolved in alcohol.

28. Birmingham University Library, Special Collections, Letter from Latham to Martineau, HM 539. Emphasis in original.

29. Birmingham University Library, Special Collections, letter from Latham to Martineau dated 18 Jan. 1855, HM 540.

30. Martineau was a regular contributor to the *Westminster Review* from the 1830s.

31. Martineau, *Autobiography*, ii. 431.

32. Harriet Martineau to Maria Weston Chapman, cited in Martineau, *Autobiography* iii. 2.

33. Wheatley, *Life*, 350.

34. Martineau, *Autobiography*, ii. 431.

35. Martineau, *Autobiography*, ii. 431.

36. Martineau, *Autobiography*, ii. 432.

37. Martineau, *Autobiography*, ii. 432.

38. Martineau, *Autobiography*, ii. 366.

39. Martineau, *Autobiography*, ii. 366.

40. Birmingham University Library, Special Collections, letter from Latham to Martineau, HM 543.

41. 'Functional' disorders contrasted to 'organic' or 'structural' complaints in that they were diseases of function rather than form. Sufferers might experience a series of physical symptoms, such as palpitations or irregular

breathing, which might be related to psychological or physical disorders. For a more detailed explanation, see Chapter 4.

42. Birmingham University Library, Special Collections, letter from Latham to Maria Martineau, HM 545.

43. See Chapter 3.

44. A condition characterized by the accumulation of serous fluid in the body's tissues.

45. Birmingham University Library, Special Collections, letter from Latham to Martineau, HM 548.

46. Sir Thomas Watson, 'Correspondence: The Late Miss Harriet Martineau', *British Medical Journal*, 8 July 1876, 64.

47. Watson, 'Correspondence', 64.

48. Watson, 'Correspondence', 64.

49. Watson, 'Correspondence', 64.

50. Watson, 'Correspondence', 64.

51. T. Spencer Wells, 'Remarks on the Case of Miss Martineau', *British Medical Journal*, 5 May 1877, 543.

52. Thomas M. Greenhow, 'Termination of the Case of Miss Harriet Martineau', *British Medical Journal*, 14 Apr. 1877, 449.

53. See the discussion in Greenhow, 'Termination', 449–50.

54. See Cooter, 'Dichotomy and Denial', 170 n. 49.

55. Ryall, 'Medical Body and Lived Experience'.

56. See Cooter, 'Dichotomy and Denial'. On the gendering of Martineau's illness, see Diana Postlethwaite, 'Mothering and Mesmerism in the Life of Harriet Martineau', *Signs*, 14 (1989), 583–609, esp. 588; Bohrer, 'Harriet Martineau: Gender, Liability and Disability', 22.

57. *Observations on those Diseases of Females which are Attended by Discharges* (London: Longman, 1814), 2; cited by Cooter, 'Dichotomy and Denial', 159.

58. See Chapter 1.

59. Blair, *Victorian Poetry*, 2.

60. Elizabeth Barrett Browning, *Sonnets from the Portuguese: A Fascimile Edition of the British Library Manuscript*, ed. William S. Paterson (1850; Barre, MA: Barre Pub.; New York: Crown Publishers, 1977); Alexander Mackenzie and Christina Rossetti, 'A Birthday ("My heart is like a singing bird")' No. 3 of 'Three Songs', *The Poetry Written by Christina Rosetti; The Music Composed by A. C. Mackenzie* (London: Novello [1878] [CPM]). Harriet Martineau, *Deerbrook* (3 vols; London: Edward Moxon, 1839; London: Virago Press, 1983).

61. Martineau, *Deerbrook*, chs 2, 3, 5.

62. Martineau, *Deerbrook*, ch. 36.

63. Martineau, *Deerbrook*, ch. 38.

64. Martineau, *Deerbrook*, chs 32–3.

65. Martineau, *Deerbrook*, ch. 22.

66. Martineau, *Deerbrook*, ch. 24.
67. Martineau, *Deerbrook*, chs 7–16; for an example of the heart beating at the arrival of a letter, see ch. 27.
68. Martineau, *Deerbrook*, ch. 36.
69. Martineau, *Deerbrook*, ch. 21.
70. Martineau, *Deerbrook*, ch. 23.
71. Martineau, *Deerbrook*, ch. 36.
72. There is no space here to elaborate further on Victorian emotion rhetoric, but see Gesa Stedman, *Stemming the Torrent: Expression and Control in the Victorian Discourses on Emotions, 1830–1872* (Aldershot: Ashgate, 2002), 75.
73. Blair, *Victorian Poetry*, 2.
74. Blair, *Victorian Poetry*, 2.
75. Contemporary examples include Robert Law, 'Disease of the Brain Dependent on Disease of the Heart', *Dublin Journal of Medical Science*, 17 (1840), 181–210; George Man Burrows, *On Disorders of the Cerebral Circulation; and on the Connection between Affections of the Brain and Diseases of the Heart* (Philadelphia: Lea and Blanchard, 1848); Joseph Swan, *An Essay on the Connections between the Action of the Heart and Arteries and the Function of the Nervous System* (London: Longman, Rees, Orme, Browne and Green, 1829); Byrom Bramwell, *Diseases of the Heart and Thoracic Aorta* (Edinburgh: Y. J. Pentland, 1884), 1.
76. e.g. John Conolly, John Forbes, and Alexander Tweedie (eds), *The Cyclopaedia of Practical Medicine* (3 vols; London: Sherwood, Gilbert and Piper, 1833).
77. See Christopher Lawrence, 'Ancients and Moderns: The "New Cardiology" in Britain, 1880–1930', in William F. Bynum, Christopher Lawrence, and Vivian Nutton (eds), *The Emergence of Modern Cardiology* (Medical History Supplement, no. 5; London: Wellcome Institute for the History of Medicine, 1985); Blair, *Victorian Poetry*.
78. See Chapter 3.
79. Blair, *Victorian Poetry*, 27.
80. On the interactions between writers and physicians, see Blair, *Victorian Poetry,* introduction and ch. 1.
81. See Chapter 4 for a discussion of Thomas Arnold's fatal angina pectoris.
82. For example, those reproduced in Rea, 'Matthew Arnold on Education'.
83. See Blair, *Victorian Poetry*, 35.
84. See Marjorie Stone, 'Browning, Elizabeth Barrett (1806–1861)', *DNB*; Jenny Uglow, 'Gaskell, Elizabeth Cleghorn (1810–1865)', *DNB*; Frank Prochaska, 'Carpenter, Mary (1807–1877)', *DNB*; Heloise Brown, 'Swaine, Ann Sykes (bap. 1821, d. 1883)', *DNB*; Martin Garrett, 'Mitford, Mary Russell (1787–1855)', *DNB*.
85. Martineau and Barrett Browning, cited in Winter, *Mesmerized*, 236.
86. Martineau, *Life in the Sick-Room*, 205.

87. On imagining the nervous body in Victorian writings, see Vrettos, *Somatic Fictions*, 47–8.

88. Blair, *Victorian Poetry*, 27.

89. Blair, *Victorian Poetry*, 29.

90. Joel Black, 'Scientific Models', in George Alexander Kennedy et al. (eds), *The Cambridge History of Literary Criticism: Romanticism* (Cambridge: Cambridge University Press, 1990), 115–38; L. Stephen Jacyna, 'Romantic Thought and the Origins of Cell Theory', in Andrew Cunningham and Nicholas Jardine (eds), *Romanticism and the Sciences* (Cambridge and New York: Cambridge University Press 1990), 161–8; Lynne Pearce, *Romance Writing* (Cambridge and Malden, MA: Polity, 2007).

CHAPTER 7

1. Francis Joseph Gall, *On the Functions of the Brain*, trans. Winslow Lewis (Boston: Marsh, Capen & Lyon, 1835), 268.

2. See, e.g. Sir John Carew Eccles, *Evolution of the Brain: Creation of the Self* (London: Routledge, 1989).

3. See, e.g. Charles Taylor, *Sources of the Self: Making of the Modern Identity* (Cambridge: Cambridge University Press, 1992).

4. In Edwin Clarke and L. Stephen Jacyna, *Nineteenth-Century Origins of Neuroscientific Concepts* (Berkeley and Los Angeles, and London: University of California Press, 1987), for instance, there is no indexed reference to the organ of the heart, nor any sustained analysis of how the rise of the brain and the neurosciences might have impacted on cardio-centric understandings of the mind-body relationship.

5. Balázs Gulyás, *The Brain–Mind Problem: Philosophical and Neurophysiological Approaches* (Louvain Philosophical Studies, 1; Leuven: Leuven University Press, 1987).

6. William Corp, *An Essay on the Changes Produced in the Body by Operations of the Mind* (London: James Ridgway, 1791); Lelland J. Rather, *Mind and Body in Eighteenth-Century Medicine: A Study Based on Jerome Gaub's "De Regime Mentis"* (London: Wellcome Historical Medical Library, 1965).

7. Robert M. Young, *Mind, Brain and Adaptation in the Nineteenth Century: Cerebral Localization and its Biological Context from Gall to Ferrier* (History of Neuroscience No. 3; New York and Oxford: Oxford University Press, 1990), preface, p. vii.

8. For a discussion, see Charles G. Gross, *Brain, Vision, Memory: Tales in the History of Neuroscience* (London: MIT Press, 1999).

9. Clarke and Jacyna, *Nineteenth-Century Origins*, 9.

10. Clarke and Jacyna, *Nineteenth-Century Origins*, 9, 12.

11. David Hothersall, *History of Psychology*, 4th edn. (New York: McGraw-Hill, 2004), 89–90. See Francis Joseph Gall, 'On Phrenology, the Localization of the Functions of the Brain' (1825), repr. in R. J. Herrnstein and E. G. Boring (eds), *A Source Book in the History of Psychology* (Cambridge, MA: Harvard University Press, 1965), 211–20.

12. Francis Joseph Gall, *On the Functions of the Brain*, trans. Winslow Lewis (Boston: Marsh, Capen & Lyon, 1835), 29–30, 112; Anthelme Richerand, *The Elements of Physiology*, trans. Robert Kerrison (London: John Murray, 1803); Jennifer Radden, 'Lumps and Bumps: Kantian Faculty Psychology, Phrenology, and Twentieth-Century Psychiatric Classification', *Philosophy, Psychiatry and Psychology*, 3 (1996), 1–14; Charles G. Gross, *Brain, Vision, Memory: Tales in the History of Neuroscience* (Cambridge, MA, and London: MIT Press, 1998), 54.

13. Gall, *On the Functions of the Brain*, p.268.

14. Marie François Xavier Bichat, *Physiological Researches on Life and Death*, trans. F. Gold (London: Longmans, n.d.), 62, 252.

15. For an introduction, see Stanley Finger, *Origins of Neuroscience: A History of Explorations into Brain Function* (New York and Oxford: Oxford University Press, 1994), 280.

16. See Finger, *Origins*, 266; Bichat, *Physiological Researches*, pp. 1–34; Erwin H. Ackerknecht, 'The History of the Discovery of the Vegetative (Autonomic) Nervous System', *Medical History*, 18 (1974), 1–8, esp. 3.

17. Ackerknecht, 'History', 3; W. H. Gaskell, *The Involuntary Nervous System* (London: Longmans, Greene and Co., 1920); John Newport Langley, *The Autonomic Nervous System* (Cambridge: W. Heffer and Sons, 1921).

18. As early as 1649, Descartes famously developed the concept of the reflex as a means of mechanistically explaining involuntary movements. This consequently made the participation of the will, the soul, or even consciousness unnecessary in such vital actions as the operation of the heartbeat. In Descartes's model, sensory perceptions travelled to the brain, from where they were reflected into the nerves to bring about the movement of the muscles. By the second half of the nineteenth century, the notion of the reflex had become an entrenched part of neuro-physiological thought, most demonstrably in the work of Marshall Hall. See Marshall Hall, *New Memoir on the Nervous System* (London: H. Baillière, 1841).

19. Finger, *Origins*, 40–1.

20. Young, *Mind, Brain and Adaptation*, 1.

21. For an introductory discussion, see Lois N. Magner, *A History of the Life Sciences* (New York: Marcel Dekker, 2002), 24.

22. See Robert J. Richards, *The Romantic Conception of Life: Science and Philosophy in the Age of Goethe* (Chicago: University of Chicago Press, 2002).

23. Clarke and Jacyna, *Nineteenth-Century Origins*, 81.

24. For an introduction, see http://www.heartmath.org/research/research-our-heart-brain.html, accessed 2 Jan. 2009. Doc Lew Childre and Howard Martin, with Donna Beech, *The HeartMath Solution: Proven Techniques for Developing Emotional Intelligence* (London: Piatkus, 1999). 'Heartmath' is discussed in more detail in the conclusion, above.

25. Arthur J. Moss, 'On the Innervation of the HeartBeat', *Annals of Noninvasive Electrocardiology*, 5 (2006), 290–1.

26. See Clarke and Jacyna, *Nineteenth-Century Origins*, 66.

27. Charles Bell, *Idea of a New Anatomy of the Brain: Submitted for the Observations of his Friends* (London: Strahan and Preston, 1811); François Magendie, 'Experiences sur les functions des raciness des nerfs rachidiens', *Journal de physiologie experimentale et pathologie*, 2 (1822), 276–9; Alexander Walker, *Documents and Dates of Modern Discoveries in the Nervous System* (1839; repr. Metuchen, NJ: Scarecrow Reprints, 1973).

28. William James, 'What is an Emotion?' *Mind*, 9 (1884), 188–205, at 188.

29. James, 'Emotion', 188.

30. James, 'Emotion', 189.

31. James, 'Emotion', 189–90.

32. James, 'Emotion', 191.

33. For a good introduction to the work of Mosso, and its contemporary resonance, see Otniel E. Dror, 'Fear and Loathing in the Laboratory and Clinic', in Fay Bound Alberti (ed.), *Medicine, Emotion and Disease, 1700–1950* (Basingstoke: Palgrave Macmillan, 2006), 125–43 and Dror, 'The Scientific Image of Emotion: Experience and Technologies of Inscription', *Configurations*, 7/3 (1999), 355–401.

34. James, 'Emotion', 191–2.

35. James, 'Emotion', 193–4.

36. Walter Bradford Cannon, 'The James-Lange Theory of Emotions: A Critical Examination and an Alternative Theory', *American Journal of Psychology*, 39 (1927), 106–34.

37. On Cannon's experimentation, see Otniel Dror, 'The Affect of Experiment: The Turn to Emotions in Anglo-American Physiology, 1900–1940', *Isis*, 90 (1999), 205–37, at 218. See Elin L. Wolfe, *Walter B. Cannon: Science and Society* (Boston, MA: Boston Medical Library in the Francis A. Countway Library of Medicine, 2000).

38. Walter B. Cannon, *Bodily Changes in Pain, Hunger, Fear and Rage. An Account of Recent Researches, etc.* 2nd edn (New York & London: D. Appleton & Co., 1929).

39. See Chapters 3 and 8. For Cannon's response to an alternative model of emotions—better known as the James-Lange theory—see Cannon, 'The James–Lange Theory of Emotions', 106–24.

40. This recognition was made by S. A. K. Wilson in 'Pathological Laughing and Crying', *Journal of Neurology and Psychopathology*, 4 (1924), 299–333, at 308–9.

41. Gerald L. Geison, *Michael Foster and the Cambridge School of Physiology* (Princeton: Princeton University Press, 1978).

42. See Rhodri Hayward, 'The Tortoise and the Love-Machine: Grey Walter and the Politics of Electroencephalography', *Science in Context*, 14 (2001), 615–41.

43. Charles Darwin, *Expression of the Emotions in Man and Animals* (London: John Murray, 1872; repr. ed. Paul Ekman, Oxford and New York: Oxford University Press, 1998).

44. Darwin, *Expression*, 77.

45. See Chapter 1.

46. Finger, *Origins*, 271.

47. On the complex relationship between Darwin and Freud, see Lucille B. Ritvo, *Darwin's Influence on Freud: A Tale of Two Cities* (New Haven and London: Yale University Press, 1990).

48. For a discussion of the influence of this belief on frontal-lobe investigation, see Finger, *Origins*, 271–3.

49. See, e.g. Francis Hutcheson, *On the Nature and Conduct of the Passions with Illustrations on the Moral Sense* (1728; repr. with an introduction by Andrew Ward, Manchester: Clinamen, 1999).

50. Dror, 'Fear and Loathing', and id., 'The Scientific Image'.

51. Dror, 'The Affect of Experiment', 207.

52. Wilhelm Max Wundt, *Outlines of Psychology*, trans. Charles Hubbard Judd (Leipzig and London: Williams & Norgate, 1897), 171–2.

53. Wilhelm Max Wundt, *Principles of Physiological Psychology* (1874; 5th edn, ed. E. Titchener, New York: Macmillan, 1904), 3. Hothersall, *History*, 123.

54. Hothersall, *History*, 123.

55. Wundt, *Outlines*, 173.

56. Wundt, *Outlines*, 175.

57. For an introduction, see Klaus R. Scherer and Paul Ekman (eds), *Approaches to Emotion* (Hillsdale, NJ: L. Erlbaum Associates, 1984).

58. Chandak Sengoopta, *The Most Secret Quintessence of Life: Sex, Glands and Hormones, 1850–1950* (Chicago: University of Chicago Press, 2006). See also Chapter 1.

59. In 2008, for instance, the topic of the 'Relational Heart' (the heart as physical, psychological, and symbolic centre of the body) was the topic of the British Holistic Medical Association's annual conference. See http://www.bhma.org for details, accessed 2 Jan. 2009.

60. See Chapter 1.

61. For a good introduction to the paradoxes of Victorian Psychology, see Rick Rylance, *Victorian Psychology and British Culture, 1850–1888* (Oxford and New York: Oxford University Press, 2000). See also id., 'Convex and Concave: Conceptual Boundaries in Psychology Now and Then (But Mainly Then)', *Victorian Literature and Culture*, 32 (2004), 449–62.

62. See Rylance, 'Convex and Concave', 449. See also Chapter 3.
63. For a more elaborate discussion of the perseverance of the soul between the seventeenth and twentieth centuries, see Edward S. Reed, *From Soul to Mind: The Emergence of Psychology from Erasmus Darwin to William James* (New Haven and London: Yale University Press, 1997), 4–5.
64. Thomas Dixon, *From Passions to Emotions: The Creation of a Secular Psychological Category* (Cambridge: Cambridge University Press, 2003), 21.
65. See John Deigh, 'Emotions: The Legacy of James and Freud', *International Journal of Psychoanalysis*, 82 (2000), 1247–56.
66. Erich Harth, *The Creative Loop: How the Brain Makes a Mind* (London: Penguin 1995); Londa Schiebinger, *Nature's Body: Gender in the Making of Modern Science* (New Brunswick, NJ: Rutgers University Press, 2004); Marianne van den Wijngaard, *Reinventing the Sexes: The Biomedical Construction of Femininity and Masculinity* (Bloomington, IN: Indiana University Press, 1997).
67. See http://news.bbc.co.uk/1/hi/health/1263758.stm, accessed 2 Jan. 2009.

CONCLUSION

1. John Banville, *The Untouchable* (London: Picador, 1997), 41.
2. William Clark, MD, *A Medical Dissertation Concerning the Effects of the Passions on Human Bodies* (London: M. Cooper in Pater-Noster Row, 1752), 52.
3. http://www.smm.org/heart/lessons/lesson5a.htm, accessed 2 Jan. 2007.
4. See Introduction.
5. Noga Arikha, *Passions and Tempers: A History of the Humours* (New York: Ecco, 2007).
6. See Chapter 3.
7. Jean Nicolas Corvisart, *An Essay on the Organic Diseases and Lesions of the Heart and Great Vessels*, trans. Jacob Gates (Boston: 1812; facs. repr. New York: Hafner, 1962). See Chapter 3.
8. Charles Bell, *Idea of a New Anatomy of the Brain* (1811; repr. London: Dawsons of Pall Mall, 1966), and François Magendie, *A Summary of Physiology* (Baltimore: E. J. Coale & Co., 1822).
9. Byrom Bramwell, *Diseases of the Heart and Thoracic Aorta* (Edinburgh: Y. J. Pentland, 1884), 38.
10. Karl Popper and John C. Eccles, 'The Self-Conscious Mind and the Brain', in eid. (eds), *The Self and its Brain* (London and New York: Routledge, Taylor & Francis Group, 1977), 355–76.
11. John A. Armour, 'Cardiac Neuronal Hierarchy in Health and Disease', *American Journal of Physiology, Regulatory, Integrative and Comparative Physiology*, 287 (2004), 262–71.

12. *Alternative Medicine: Expanding Medical Horizons: A Report to the National Institutes of Health on Alternative Medical Systems and Practices in the United States* (NIH publication, no. 94–066; National Institutes of Health, 1994).

13. The interplay of a variety of causal factors, from depression to increased incidence of accidents, is acknowledged in the research. Thanks to Martin Cowie for providing access to his 2007 unpublished paper, 'Don't Go Breaking My Heart: A Medical Perspective'.

Indicative Bibliography

Acierno, Louis J., *History of Cardiology* (London: Parthenon, 1993).

Ackerknecht, Erwin H., 'The History of the Discovery of the Vegetative (Autonomic) Nervous System', *Medical History*, 18 (1974), 1–8.

Arikha, Noga, *Passions and Tempers: A History of the Humours* (New York: Ecco, 2007).

Baillie, Matthew, *The Morbid Anatomy of Some of the Most Important Parts of the Human Body* (London: J. Johnson and G. Nicol, 1793; facs. repr. Birmingham, AL: Classics of Medicine Library, 1986).

Bamborough, John Bernard, *The Little World of Man* (London and New York: Longmans, Green, 1952).

Barrett, William, *Death of the Soul: From Descartes to the Computer* (New York: Doubleday, 1986).

Bean, William B., *Aphorisms from Latham* (Iowa City: Prairie Press [1962]).

Beard, George Miller, *Sexual Neurasthenia (Nervous Exhaustion): Its Hygiene, Causes, Symptoms and Treatment with a Chapter on Diet for the Nervous*, ed. A. D. Rockwell (5th edn, New York: E. B. Treat, 1898).

Bell, Charles, *Idea of a New Anatomy of the Brain: Submitted for the Observations of his Friends* (London: Strahan and Preston, 1811).

Benjamin, Marina (ed.), *Science and Sensibility: Gender and Scientific Enquiry, 1780–1945* (Oxford: Basil Blackwell, 1991).

Bichat, Marie-François-Xavier, *General Anatomy, Applied to Physiology and the Practice of Medicine*, rev. George Calvert (London: Printed for the Translator, 1824).

—— *Physiological Researches on Life and Death*, trans. F. Gold (London: Longmans, [n.d.]).

Blair, Kirstie, *Victorian Poetry and the Culture of the Heart* (Oxford: Oxford University Press, 2006).

Bohrer, Susan F., 'Harriet Martineau: Gender, Liability and Disability', *Nineteenth-Century Contexts*, 25 (2003), 21–37.

Bound, Fay, '"Writing the Self": Love and the Letter in England, *c*.1660–*c*.1760', *Literature and History*, 11 (2002), 1–19.

—— 'An "Angry and Malicious Mind"? Narratives of Slander at the Church Courts of York, *c*.1660–*c*.1760', *History Workshop Journal*, 56 (2003), 59–77.

Bound Alberti, Fay (ed.), *Medicine, Emotion and Disease, 1700–1950* (Basingstoke: Palgrave, 2006).

Bound Alberti, Fay (ed.), 'The Emotional Heart: Mind, Body and Soul', in James Peto (ed.), *The Heart* (New Haven and London: Yale University Press, 2007) 125–42.

——'Angina Pectoris and the Arnolds: Emotions and Heart Disease in Nineteenth-Century Culture', *Medical History*, 52 (2008), 221–36.

Bourke, Joanna, *Fear: A Cultural History* (London: Virago Press, 2005).

Boyadjian, Noubar, *The Heart: Its History, its Symbolism, its Iconography and its Diseases* (Antwerp: Esco, 1980).

Bramwell, Byrom, *Diseases of the Heart and Thoracic Aorta* (Edinburgh: Y. J. Pentland, 1884).

Buchan, William, *Domestic Medicine: or, a Treatise on the Prevention and Cure of Diseases by Regimen and Simple Medicines* (1772 edn; facs. repr. New York and London: Garland, 1985).

Burkitt, Ian, *Bodies of Thought: Embodiment, Identity and Modernity* (London: Sage, 1999).

——*Social Selves: Theories of Self and Society* (2nd edn, Los Angeles and London: Sage, 2008).

Burns, Allan, *Observations on Some of the Most Frequent and Important Diseases of the Heart: On Aneurism of the Thoracic Aorta; On Preternatural Pulsation in the Epigastric Region* (Edinburgh: Muirhead, 1809).

Burrows, George Man, *On Disorders of the Cerebral Circulation; and on the Connection between Affections of the Brain and Diseases of the Heart* (Philadelphia: Lea and Blanchard, 1848).

Burton, Robert, *Anatomy of Melancholy* (1621; repr. New York: New York Review of Books, 2001).

Butter, William, *A Treatise on the Disease Commonly Called Angina Pectoris* (London, 1791).

Bynum, William F., Lawrence, Christopher, and Nutton, Vivian (eds), *The Emergence of Modern Cardiology* (London: Wellcome Institute for the History of Medicine, 1985).

Cannon, Walter B., *The James-Lange Theory of Emotions: A Critical Examination and an Alternative Theory* (Ithaca, NY: American Journal of Psychology, 1927).

——*Bodily Changes in Pain, Hunger, Fear and Rage. An Account of Recent Researches, etc.* (2nd edn, New York and London: D. Appleton & Co., 1929).

Charleton, Walter, *Natural History of the Passions* ([London] In the Savoy: Printed by T. N. for J. Magnes, 1674).

Childre, Doc Lew, and Martin, Howard, with Beech, Donna, *The HeartMath Solution: Proven Techniques for Developing Emotional Intelligence* (London: Piatkus, 1999).

Churchland, Patricia S., *Neurophysiology: Toward a Unified Science of the Mind/Brain* (Cambridge, MA: MIT Press, 1986).

Clark, William, *A Medical Dissertation Concerning the Effects of the Passions on Human Bodies* (Bath and London: for W. Frederick, 1752).

Clarke, Edwin, and Jacyna, L. Stephen *Nineteenth-Century Origins of Neuro-scientific Concepts* (Berkeley and Los Angeles, and London: University of California Press, 1987).

Cogan, Thomas, *A Treatise on the Passions and Affections of the Mind* (5 vols; London: T. Cadell and W. Davies, 1813).

Coleman, William, and Holmes, Frederic L. (eds), *The Investigative Enterprise* (Berkeley and Los Angeles: University of California Press, 1988).

Cook, Harold J., 'Boerhaave and the Flight from Reason in Medicine', *Bulletin of the History of Medicine*, 74 (2000), 221–40.

Cooter, Roger, *The Cultural Meaning of Popular Science: Phrenology and the Organization of Consent in Nineteenth-Century Britain* (Cambridge: Cambridge University Press, 1984).

Corvisart, Jean-Nicolas, *An Essay on the Organic Diseases and Lesions of the Heart and Great Vessels*, trans. Jacob Gates (Boston: 1812; facs. repr. New York: Hafner, 1962).

Crooke, Helkiah, *Mikrokosmographia: A Description of the Body of Man* (London, 1615).

Cullen, William, *Synopsis Nosologiae Methodicae* (Edinburgh, 1769).

Cunningham, Andrew, and French, Roger Kenneth (eds), *The Medical Enlightenment of the Eighteenth Century* (Cambridge and New York: Cambridge University Press, 1990).

—— and Jardine, Nicholas (eds), *Romanticism and the Sciences* (Cambridge and New York: Cambridge University Press, 1990).

Damasio, Antonio, *The Feeling of What Happens: Body, Emotion and the Making of Consciousness* (New York: Harcourt, 2000).

Darwin, Charles, *Expression of the Emotions in Man and Animals* (London: John Murray, 1872; repr. ed. Paul Ekman, Oxford and New York: Oxford University Press, 1998).

Daston, Lorraine, and Galison, Peter, *Objectivity* (New York: Zone, 2007).

Davies, Herbert, *Lectures on the Physical Diagnosis of the Diseases of the Lungs and Heart* (London: John Churchill, 1851).

Descartes, René, *Les passions de l'âme* (Amsterdam: [Henry Le Gras] chez Louys Elzevier, 1649)

Dixon, Thomas, *From Passions to Emotions: The Creation of a Secular Psychological Category* (Cambridge: Cambridge University Press, 2003).

Doueihi, Milad, *A Perverse History of the Human Heart* (Cambridge, MA, and London: Harvard University Press, 1997).

Downame, John, *A Treatise of Anger* (London: Printed by T. E. for William Welby, 1609).

Dror, Otniel, 'The Affect of Experiment: The Turn to Emotions in Anglo-American Physiology, 1900–1940', *Isis*, 90 (1999), 205–37.

—— 'The Scientific Image of Emotion: Experience and Technologies of Inscription', *Configurations*, 7 (1999), 355–401.

Duffin, Jacalyn, *To See with a Better Eye: A Life of R. T. H. Laennec* (Princeton: Princeton University Press, 1998).

East, Terence, *The Story of Heart Disease* (London: William Dawson, 1958).

Eccles, John Carew, *Evolution of the Brain: Creation of the Self* (London: Routledge, 1989).

Ekman, Paul, *The Nature of Emotion: Fundamental Questions* (New York: Oxford University Press, 1995).

Erickson, Robert A., *The Language of the Heart, 1600–1750* (Philadelphia: University of Pennsylvania Press, 1997).

Finger, Stanley, *Origins of Neuroscience: A History of Explorations into Brain Function* (New York and Oxford: Oxford University Press, 1994).

Fleming, Peter R., *A Short History of Cardiology* (Amsterdam: Rodopi, 1997).

Forth, Christopher E., and Carden-Coyne, Ana (eds), *Cultures of the Abdomen: Diet, Digestion and Fat in the Modern World* (Basingstoke: Palgrave Macmillan, 2005).

Fothergill, John, 'Farther Account of the Angina Pectoris', *Medical Observations and Inquiries by a Society of Physicians in London*, 5 (1776), 252–8.

Frawley, Maria, 'Harriet Martineau: Health and Journalism', *Women's Writing*, 9 (2002), 433–44.

French, Roger Kenneth, *The History of the Heart: Thoracic Physiology from Ancient to Modern Times* (Aberdeen: Equipress, 1979).

——— and Wear, Andrew (eds), *The Medical Revolution of the Seventeenth Century* (Cambridge: Cambridge University Press, 1989).

——— (eds), *British Medicine in an Age of Reform* (New York: Routledge, 1991).

Fye, Bruce W., *American Cardiology: The History of a Specialty and its College* (Baltimore and London: Johns Hopkins University Press, 1996).

Galen, *On the Passions and Errors of the Soul*, trans. Paul W. Harkins ([Columbus]: Ohio State University Press, 1963).

Gall, Francis Joseph, *On the Functions of the Brain*, trans. Winslow Lewis (Boston: Marsh, Capen & Lyon, 1835).

Geison, Gerald L., *Michael Foster and the Cambridge School of Physiology: The Scientific Enterprise in Late Victorian Society* (Princeton: Princeton University Press, 1978).

Gikswijt-Hofstra, Marijke, and Porter, Roy (eds), *Cultures of Neurasthenia from Beard to the First World War* (Wellcome Institute Series in the History of Medicine, 63; Amsterdam: Rodopi, 2001).

Gregory, Andrew, *Harvey's Heart: The Discovery of Blood Circulation* (London: Icon, 2001).

Gross, Charles G., *Brain, Vision, Memory: Tales in the History of Neuroscience* (Cambridge, MA, and London: MIT Press, 1998).

Haigh, Elizabeth L., 'Vitalism, the Soul and Sensibility: The Physiology of Théophile Bordeu', *Journal of the History of Medicine and Allied Sciences*, 1 (1976), 30–41.

Halliwell, Martin, *Romantic Science and the Experience of Self: Transatlantic Crosscurrents from William James to Oliver Sacks* (Aldershot and Brookfield, VT: Ashgate, 1999).

Harvey, William, *Exercitatio Anatomica de Motu Cordis et Sanguinus in Animalibus* (Frankfurt: Sumptibus Guilielmi Fitzeri, 1628).

Helmont, Jean-Baptiste van, *The Spirit of Diseases; or, Diseases from the Spirit... Wherein is Shewed how Much the Mind Influenceth the Body in Causing and Curing of Diseases* (London: Sarah Howkins, 1694).

Hobbes, Thomas, *Leviathan*, ed. Richard Tuck (1651; repr. Cambridge: Cambridge University Press, 1991).

Hope, James, *A Treatise on the Diseases of the Heart and the Great Vessels Comprising a New View of the Physiology of the Heart's Action* (London: William Kidd; Edinburgh: Adam Black; Glasgow: T. Atkinson & Co., 1832).

Hothersall, David *History of Psychology* (4th edn, New York: McGraw-Hill, 2004).

Høystad, Ole Martin, *A History of the Heart* (London: Reaktion, 2007).

Huisman, Frank, and Warner, John Harley (eds) *Locating Medical History: The Stories and their Meanings* (Baltimore and London: Johns Hopkins University Press, 2006).

Hunter, John, *Treatise on Venereal Disease*, 2nd edn, trans. and ed. Freeman J. Bumstead (Philadelphia: Blanchard and Lea, 1859).

—— *The Works of John Hunter, F.R.S.*, ed. James F. Palmer (4 vols; London: Longman, Rees, Orme, Browne, Green and Longman, 1835).

Jackson, Stanley W., *Melancholia and Depression: From Hippocratic Times to Modern Times* (New Haven and London: Yale University Press, 1986).

Jager, Eric, *The Book of the Heart* (Chicago and London: University of Chicago Press, 2000).

James, Susan, *Passion and Action: The Emotions in Seventeenth-Century Philosophy* (Oxford: Oxford University Press, 1997).

James, William, 'What is an Emotion?' *Mind*, 9 (1884), 188–205.

Julian, Desmond Gareth (ed.), *Diseases of the Heart* (London: Saunders, 1996).

Kuriyama, Shigehisa, *The Expressiveness of the Body and the Divergence of Greek and Chinese Medicine* (New York: Zone Books, 1999).

Laennec, Rene T. H., *A Treatise on the Diseases of the Chest*, trans. John Forbes (London: 1821; facs. New York: Hafner Publishing, 1962).

—— *A Treatise on the Diseases of the Chest and on Mediate Auscultation*, 2nd edn, trans. John Forbes (London: Thomas and George Underwood, 1827).

Lawrence, Christopher, 'Incommunicable Knowledge: Science, Technology and the Clinical Art in Britain 1850–1914', *Journal of Contemporary History*, 20 (1985), 503–20.

Lawrence, Christopher, and Weisz, George (eds), *Greater than the Parts: Holism in BioMedicine, 1920–1950* (New York and Oxford: Oxford University Press, 1998).

Latham, Peter Mere, *Lectures on Subjects Connected with Clinical Medicine, Comprising Diseases of the Heart* (2 vols; London: Longman, Brown, Green and Longmans, 1845).

——'General Remarks on the Practice of Medicine: The Heart and its Affections, Not Organic', *British Medical Journal*, 1 (1863), 1–3.

Leibowitz, Joshua O., *The History of Coronary Heart Disease* (London: Wellcome Institute, 1970).

Lettsom, John Coakley, *The Works of John Fothergill* (London: Charles Dilly, 1873).

Livesley, Brian, 'The Spasms of John Hunter: A New Interpretation', *Medical History*, 17 (1973), 70–5.

Logan, Deborah Ann, *The Hour and the Woman: Harriet Martineau's 'Somewhat Remarkable Life* (Dekalb, IL: Northern Illinois University Press, 2002).

Lüderitz, Berndt, *History of the Disorders of Cardiac Rhythm* (Armonk, NY: Futura Pub. Co., 2002).

Macdonald, Paul S., *History of the Concept of Mind: Speculations about Soul, Mind and Spirit from Homer to Hume* (Aldershot: Ashgate, 2007).

McGann, Jerome, *The Poetics of Sensibility: A Revolution in Literary Style* (Oxford: Oxford University Press, 1998).

Mackenzie, James, *The Study of the Pulse: Arterial, Venous, and Hepatic, and of the Movements of the Heart* (Edinburgh and London: Young J. Pentland, 1902).

Magendie, François, *A Summary of Physiology* (Baltimore: E. J. Coale & Co., 1822).

Martineau, Harriet, *Miss Martineau's Letters on Mesmerism* (New York: Harper and Brothers, 1845).

——*Harriet Martineau's Autobiography*, intro. Gaby Weiner (3 vols; London: 1877; London: Virago, 1983).

——*Deerbrook* (3 vols; London: Edward Moxon, 1839; London: Virago Press, 1983).

——*Life in the Sickroom*, ed. Maria H. Frawley (Ormskirk: Broadview, 2003).

Maulitz, Russell C., *Morbid Appearances: The Anatomy of Pathology in the Early Nineteenth Century* (Cambridge and New York: Cambridge University Press, 1997).

Mayr, Otto, *Authority, Liberty and Automatic Machines in Early Modern Europe* (Baltimore: Johns Hopkins University Press, 1986).

Moore, Wendy, *The Knife Man: The Extraordinary Life and Times of John Hunter, Father of Modern Surgery* (London: Bantam, 2005).

Osler, William, *The Principles and Practice of Medicine* (Edinburgh and London: Young J. Pentland, 1898).

Parry, Caleb Hillier, *An Inquiry into the Symptoms and Causes of the Syncope Anginosa, Commonly Called Angina Pectoris* (London: R. Crutwell, 1799).

Paster, Gail Kern, Rowe, Katherine, and Floyd-Wilson, Mary (eds), *Reading the Early Modern Passions: Essays in the Cultural History of Emotion* (Philadelphia: University of Philadelphia Press, 2004).

Peto, James (ed.), *The Heart* (New Haven and London: Yale University Press, 2007).

Pettigrew, Thomas J., 'John Hunter: From the Medical Portrait Gallery', *Lancet*, 815 (1939), 119–20.

Pickstone, John V., *Ways of Knowing: A New History of Science, Technology and Medicine* (Manchester: Manchester University Press, 2000).

Popper, Karl, and John C. Eccles (eds), *The Self and its Brain* (London and New York: Routledge, Taylor & Francis Group, 1977).

Porter, Roy, *Enlightenment: Britain and the Creation of the Modern World* (London: Allen Lane, 2000).

Rather, Lelland Joseph, 'Old and New Views of the Emotions and Bodily Changes: Wright and Harvey versus Descartes, James and Cannon', *Clio Medica*, 1 (1965), 1–25.

Reddy, William, *Navigation of Feeling: A Framework for the History of Emotions* (Cambridge: Cambridge University Press, 2001).

Reed, Edward S., *From Soul to Mind: The Emergence of Psychology from Erasmus Darwin to William James* (New Haven and London: Yale University Press, 1997).

Rhawn, Joseph, *Neuropsychiatry, Neuropsychology, and Clinical Neuroscience: Emotion, Evolution, Cognition, Language, Memory, Brain Damage, and Abnormal Behaviour* (2nd edn, Baltimore: Williams and Wilkins, 1996).

Richardson, Robert Galloway, *Heart and Scalpel: A History of Cardiac Surgery* (London: Quiller Press, 2001).

Rimé, Bernard, 'The Social Sharing of Emotion as an Interface between Individual and Collective Processes in the Construction of Emotional Climates', *Journal of Social Issues*, 63 (2007), 307–22.

Roberts, Caroline, *The Woman and the Hour: Harriet Martineau and Victorian Ideologies* (Toronto: University of Toronto Press, 2002).

Rodin, Alvin E., *The Influence of Matthew Baillie's Morbid Anatomy: Biography, Evaluation and Reprint* (Springfield, IL: Charles C. Thomas, 1973).

Rosenberg, Charles E., and Janet Golden (eds), *Framing Disease: Studies in Cultural History* (New Brunswick, NJ: Rutgers University Press, 1992).

Rosenwein, Barbara H., 'Worrying about Emotions in History', *American Historical Review*, 107 (2002), 821–45.

——*Emotional Communities in the Early Middle Ages* (Ithaca, NY: Cornell, 2007).

Rousseau, George S. (ed.), *The Languages of Psyche: Mind and Body in Enlightenment Thought* (Berkeley and Los Angeles, and Oxford: University of California Press, 1990).

——*Enlightenment Borders: Pre- and Post-Modern Discourses, Medical and Scientific* (Manchester: Manchester University Press, 1991).

——*Nervous Acts: Essays on Literature, Culture and Sensibility* (Basingstoke and New York: Palgrave Macmillan, 2004).

Ryall, Anka, 'Medical Body and Lived Experience: The Case of Harriet Martineau', *Mosaic*, 33 (2000), 35–53.

Rylance, Rick, *Victorian Psychology and British Culture, 1850–1888* (Oxford and New York: Oxford University Press, 2000).

—— 'Convex and Concave: Conceptual Boundaries in Psychology Now and Then (but Mainly Then)', *Victorian Literature and Culture*, 32 (2004), 449–62.

Sanders, Valerie, *The Private Lives of Victorian Women: Autobiography in Nineteenth-Century England* (New York and London: Harvester Wheatsheaf, 1989).

Scherer, Klaus R., and Paul Ekman (eds), *Approaches to Emotion* (Hillsdale, NJ: L. Erlbaum Associates, 1984).

Schore, Allan N., *Affect Regulation and the Origin of the Self: The Neurobiology of Emotional Development* (Hillsdale, NJ, and Hove: L. Erlbaum Associates, 1994).

Sengoopta, Chandak, *The Most Secret Quintessence of Life: Sex, Glands, and Hormones, 1850–1950* (Chicago: University of Chicago Press, 2006).

Shorter, Edward, *From Paralysis to Fatigue: A History of Psychosomatic Illness in the Modern Era* (New York: Free Press, 1992).

Shumacker, Harris B., *The Evolution of Cardiac Surgery* (Bloomington, IN: Indiana University Press, 1992).

Snellen, H. A., *History of Cardiology: A Brief Outline of the 350 Years' Prelude to an Explosive Growth* (Rotterdam: Donker Academic Publications, 1984).

Stanley, Arthur Penrhyn, *Life of Thomas Arnold, D.D. Headmaster of Rugby* (1844; repr. London: John Murray, 1904).

Stansfeld, Stephen, and Marmot, Michael (eds), *Stress and the Heart: Psychosocial Pathways to Coronary Heart Disease* (London: BMJ Books, 2002).

Stearns, Peter N., *Jealousy: The Evolution of an Emotion in American History* (New York and London: New York University Press, 1989).

—— *Battleground of Desire: The Struggle for Self-Control in Modern America* (New York and London: New York University Press, 1999).

—— and Stearns, Carol Z., 'Emotionology: Clarifying the History of Emotions and Emotional Standards', *American Historical Review*, 90 (1985), 813–36.

———— *Anger: The Struggle for Emotional Control in America's History* (Chicago and London: University of Chicago Press, 1986).

Stedman, Gesa, *Stemming the Torrent: Expression and Control in the Victorian Discourses on Emotions, 1830–1872* (Aldershot: Ashgate, 2002).

Swan, Joseph, *An Essay on the Connections between the Action of the Heart and Arteries and the Function of the Nervous System* (London: Longman, Rees, Orme, Browne and Green, 1829).

Taylor, Charles, *Sources of the Self: Making of the Modern Identity* (Cambridge: Cambridge University Press, 1992).

Temkin, Oswei, *Galenism: Rise and Decline of a Medical Philosophy* (Ithaca, NY, and London: Cornell University Press, 1973).

Trevor, Meriol, *The Arnolds: Thomas Arnold and his Family* (London: Bodley Head, 1973).

Veith, Ilza, *Hysteria: The History of a Disease* (Chicago: University of Chicago Press, 1970).

Vincent-Buffault, Anne, *A History of Tears: Sensibility and Sentimentality in France*, trans. Teresa Bridgeman (Basingstoke: Macmillan, 1990).

Vogel, John H. K., and Krucoff, Mitchell W. (eds), *Integrative Cardiology: Complementary and Alternative Medicine for the Heart* (New York: McGraw-Hill Medical; London: McGraw Hill [distributor], 2007).

Von Haller, Albrecht, *First Lines of Physiology*, ed. Lester S. King (2 vols; 1786, facs. repr. New York: Johnson Reprint Corp., 1966).

Vrettos, Athena, *Somatic Fictions: Imagining Illness in Victorian Culture* (Stanford, CA: Stanford University Press, 1995).

Warner, John Harley, *The Therapeutic Perspective: Medical Practice, Knowledge, and Identity in America, 1820–1885* (Cambridge, MA, and London: Harvard University Press, 1986).

—— *Against the Spirit of System: The French Impulse in Nineteenth-Century American Medicine* (Baltimore: Johns Hopkins University Press, 2003).

Warren, John Collins, *Cases of Organic Diseases of the Heart. With Dissections and Some Remarks Intended to Point out the Distinctive Symptoms of these Diseases* (Boston: Thomas B. Wait and Company, 1809).

Wear, Andrew, *Medicine in Society: Historical Essays* (Cambridge: Cambridge University Press, 1992).

Weidner, Gerdi, Kopp, Mária, and Kristenson, Margareta (eds), *Heart Disease: Environment, Stress and Gender* (NATO Science Series, Series I, vol. 327; Budapest: IOS, 2000).

Whitaker, Harry, Christopher Upham Murray Smith and Stanley Finger (eds), *Brain, Mind and Medicine: Essays in Eighteenth-Century Neuroscience* (New York: Springer, 2007).

Wierzbicka, Anna, *Emotions across Languages and Cultures: Diversity and Universals* (Cambridge: Cambridge University Press, 1999).

Williams, Simon J., 'Modernity and the Emotions: Corporeal Reflections on the (Ir)rational', *Sociology*, 32 (1988), 747–69.

Williamson, James M., *Ventnor and the Undercliff in Chronic Pulmonary Diseases* (London: Bailliere, Tindall and Cox, 1884).

Willis, Thomas, *Dr Willis's Practice of Physick* (London: T. Dring, C. Harper, and J. Leigh, 1684).

Willius, Frederick A., and Keys, Thomas (eds), *Cardiac Classics* (London: Henry Kipton, 1941).

Winter, Alison, *Mesmerized: Powers of Mind in Victorian Britain* (Chicago and London: University of Chicago Press, 1998).

Wooley, Charles F., *The Irritable Heart of Soldiers and the Origins of Anglo-American Cardiology: The US Civil War (1861) to World War I (1918)* (Aldershot: Ashgate, 2002).

Wright, John P., and Potter, Paul (eds), *Psyche and Soma: Physicians and Metaphysicians on the Mind–Body Problem from Antiquity to Enlightenment* (Oxford: Oxford University Press, 2000).

Wright, Thomas, *Passions of the Minde in Generall* (1601, 1604; repr. Urbana, IL: University of Illinois Press, 1971).

Wundt, Wilhelm Max, *Outlines of Psychology*, trans. Charles Hubbard Judd (Leipzig and London: Williams & Norgate, 1897).

—— *Principles of Physiological Psychology*, 5th edn, ed. E. Titchener (1874; New York: Macmillan, 1904).

Young, Louisa, *The Book of the Heart* (London: Flamingo, 2002).

Young, Robert M., *Mind, Brain and Adaptation in the Nineteenth Century: Cerebral Localization and its Biological Context from Gall to Ferrier* (History of Neuroscience, 3; New York and Oxford: Oxford University Press, 1990).

Index